AN INVARIANT APPROACH
TO
STATISTICAL
ANALYSIS OF SHAPES

CHAPMAN & HALL/CRC
Interdisciplinary Statistics Series

Series editors: N Keiding, B Morgan, P van der Heijden.

AN INVARIANT APPROACH TO
STATISTICAL ANALYSIS OF SHAPES

Interdisciplinary Statistics

Subhash R. Lele

Department of Mathematical Sciences
University of Alberta
Edmonton, Canada

Joan T. Richtsmeier

Department of Anthropology
The Pennsylvania State University
University Park, Pennsylvania, USA

CRC Press

Taylor & Francis Group
Boca Raton London New York

CRC Press is an imprint of the
Taylor & Francis Group, an **informa** business

A CHAPMAN & HALL BOOK

CRC Press
Taylor & Francis Group
6000 Broken Sound Parkway NW, Suite 300
Boca Raton, FL 33487-2742

First issued in paperback 2019

ISBN-13: 978-0-8493-0319-7 (hbk)
ISBN-13: 978-0-367-39763-0 (pbk)
Library of Congress Card Number 00-050845

Library of Congress Cataloging-in-Publication Data

Lele, Subhash
    An invariant approach to statistical analysis of shapes/Subhash R. Lele, Joan T. Richtsmeier.
        p. cm. — (Interdisciplinary statistics)
    Includes bibliographical references and index.
    ISBN 0-8493-0319-2 (alk. paper)
    1. Morphology—Statistical methods. 2. Morphology—Mathematics. I. Richtsmeier,
Joan T. II. Title III. Series.

QH351 .L46 2000
571.3'01'5195—dc21

                                      00-050845
                                        CIP

Visit the Taylor & Francis Web site at
http://www.taylorandfrancis.com

and the CRC Press Web site at
http://www.crcpress.com

Dedicated to the memory of our mothers,
Sarojini Lele and Mary Hill Richtsmeier

Contents

Preface

"Faith is the bird that feels the light and sings while the dawn is still dark"

Rabindranath Tagore

Quantitative study of form and form change comprises the field of morphometrics. Cuvier (1828) was one of the first biologists to verbalize the dictum "form follows function". Charles Darwin's work on the theory of natural selection and evolution relied heavily on the study of form and especially variation in form (Darwin, 1859). The seminal work of D'Arcy Thompson (Thompson, 1917) formulated the subject in detail. Substantial developments in both biological and statistical aspects of morphometrics occurred over the next several decades of the twentieth century. Work by Mahalanobis, Rao, and their colleagues initiated the use of multivariate statistical analysis for classification of organisms into groups. Julian Huxley (Huxley, 1932) formulated the field of allometry studying the relationship between size and shape of organisms. James Mosimann (1970) constructed a proper statistical foundation for the ideas of size, shape and allometry.

The method of superimposition, particularly the Procrustes superimposition, was developed and introduced to the biological sciences by the famed anthropologist Franz Boaz and his student Eleanor Phelps (Boas, 1905; Phelps, 1932; see Cole, 1996). Later, Sneath (1967) initiated the use of explicit deformation functions for modeling form change. In the last two decades, the idea of studying form change using superimposition and deformation approaches has been seriously considered and further developed by several individuals. While Bookstein considered the deformation approach, Kendall and his colleagues Mardia, Goodall, Small, and others concentrated on superimposition techniques. A particular deformation approach, Finite Element Scaling Analysis, was developed by bioengineers (Lew and Lewis, 1977; Lewis et al., 1980) and then applied to additional biological problems by Cheverud and his colleagues (Cheverud, et al., 1983, 1991; Richtsmeier and Cheverud, 1986). However, finite element scaling analysis was never fully embraced by biologists. Some of the reluctance felt by biologists stemmed from the seemingly complex mathematics that served as the foundation of the finite element method, but the lack of invariance of this method and other superimposition techniques was recognized (Moyers and Bookstein, 1982; Cheverud and Richtsmeier, 1987; Richtsmeier, 1990). Lele (1991) formalized a precise statement regarding the lack of invariance in morphometrics and provided the solution that is invariant to the arbitrary choice of coordinate system. This monograph summarizes and synthesizes the development of this solution in the context of significant scientific problems.

This work is a collaborative effort between a statistician (SL) and a biologist (JTR), each one making the other think more deeply and carefully about the problems and solutions. It is intended for both biologists and statisticians. We have strived to make discussions as mathematically and statistically precise as possible, while keeping "the science," that is the scientific question posed at the top of our agenda.

We feel it necessary to caution the reader on one aspect of this book: this is not a statistics textbook. Although some of the basic concepts in statistics are explained in the text at an intuitive level, these discussions are not, in any way, meant as a substitute or replacement for the study of a proper statistical textbook. Most chapters have two parts. Part 1 is intended to be accessible to biologists with some statistical training, thus making it more intuitive and unfortunately, somewhat more vague and less mathematically rigorous. The underlying mathematical concepts require equations. Part 2 of most chapters contains fully rigorous mathematical arguments. Critical readers of statistical methodology should concentrate on the part 2 of most chapters. We have chosen pedagogy over mathematical rigor in Part 1 of each chapter, knowing that mathematical rigor will be demonstrated properly in Part 2. We also have provided fairly detailed computational algorithms, when appropriate.

This book is composed of seven chapters. Chapters 2, 3, and 4 have two parts. Chapters 1, 2, 5 and 6 are written to be accessible to all readers. Part 1 of Chapters 3 to 5 contain notation and mathematical concepts, but are written to be accessible to the quantitative biologist. Part 2 of Chapters 3 and 4 are targeted towards statisticians, or more advanced quantitative biologists. Included in Chapters 2 through 6 are detailed computational algorithms for the implementation of various methods. These are targeted towards statisticians, or more advanced quantitative biologists. The book is organized in this way so that the more difficult mathematical portions can be passed over without loss of continuity or of

Acknowledgments

This work could never have been done without the help, advice and contributions of many individuals. First and foremost, we would like to express our gratitude to Stephanie Harding and Sharon Taylor, our editors, for keeping the faith and trusting that we would indeed finish the project while at the same time prodding us to actually "get the book done." Their patience and trust will forever be appreciated. This project started under the editorship of John Kimmel and then was taken over by Mark Pollard, both of whom were extremely helpful in their roles as editors. We are also grateful to Bill Heyward Bob Stern, Maggie Mogck, and the production staff for their help and patience.

There were several statistician colleagues whose encouragement and thoughts have helped us along. We would like to thank Noel Cressie, Bruce Lindsay, Charles E. McCulloch, George Casella, Yehuda Vardi, C. R. Rao, and an anonymous Associate Editor for the *Journal of the American Statistical Association* who asked the most insightful questions. We would like to thank our collaborators from the fields of Neurosurgery and Reconstructive and Plastic Surgery whose advice and generosity enabled the application of our methods to morphometric problems of true consequence. These colleagues include: Dr. Ben Carson, Dr. Alex Kane, Dr. Jeffrey Marsh, and Dr. Craig VanderKolk. Encouragement and penetrating questions from our biologist colleagues were also instrumental throughout the development of our methodology. We specifically thank Jim Cheverud, Bill Atchley, Roger Reeves, Craig VanderKolk, Tim Cole, Alex Kane, Brian Corner, Norm McLeod, J.D. Singh, and Bill Jungers. Students, post docs and technologists of the Richtsmeier lab who suffered through the initial drafts of this monograph providing invaluable advice include: Anita Lubensky, Yizheng Li, Michael Zumpano, Kristina Aldridge, Jay Mussell, Frank Williams, Jideotor Aniukwu, Gail Krovitz, Valerie DeLeon, and Christopher Valeri. Bruce Latimer, Rich Sherwood, and Dana Duren took time from their busy schedules to provide photographs of the monkey skulls that were the basis for several figures in the book. A special thanks to Peter Elfert for technological assistance with the transfer of computed tomography scans to the Richtsmeier lab. We want to congratulate and thank Tim Phelps, the main artist for the book, for executing most of the figures. Valerie Burke DeLeon created the remaining figures and we are grateful for her artistic talents, as well as for her sharp editorial eye. Kristina Aldridge and Valerie DeLeon designed the cover illustration.

Anytime the establishment is challenged, there is inevitable controversy. We would specifically like to thank Michael Douglas, Matt Cartmill, Jim Cheverud, Robert Ehrlich, and George Casella, the editors of some of the journals where our original papers were published. These individuals stood firm against inappropriate comments and reviews. Animosity can occur when individuals have a passion for their work. We appreciate the efforts of Colin Goodall, Fred Bookstein, James Rohlf, and colleagues who have helped us strengthen our defenses and sharpen our methods.

Our research was supported by various agencies and private contributors including: the National Science Foundation, the National Institutes of Health, the March of Dimes Foundation, The Whitaker Foundation, Johns Hopkins University Biomedical Research Support grants, Storz Instrument Company, Ungermann-Bass, Inc., Synexus, Inc., and the Department of the Army. We gratefully acknowledge their support. We would also like to thank members of the faculty of the Department of Biostatistics and the Department of Cell Biology and Anatomy at the Johns Hopkins University for providing a stimulating atmosphere where most of the work summarized in this book was done. If there are any corrections or changes in the content of the book, they will be available on http://faith.med.jhmi.edu.

CHAPTER 1

Introduction

"Nor must we forget that the biologist is much more exacting in his requirements, as regards form, than the physicist; for the latter is usually content with either an ideal or a general description of form, while the student of living things must be specific."

D'Arcy Thompson (1992)

Natural scientists have long perceived and classified organisms primarily on the basis of their appearance and structure, otherwise known as their *form*. Beyond classification, students of biology study form in order to understand those processes that underlie variation in form, processes like disease, growth, and evolution. The true form of an object does not change whether an organism is moving across a surface (translation in mathematical terms) or spinning on an axis (rotation). Our perception of the organism may change when orientation changes, but perception is not our concern here. It follows that any quantitative representation of a form should not change if the coordinate system used to represent that form changes. Moreover, any comparison of forms should give us equivalent results regardless of whether the operations of translation and rotation are used in these comparisons. These simple observations underscore the importance of the principle of *invariance* in the study of form.

The invariance principle and its implications for the study of biological form have prompted us to write this book. There are now many morphometric methods available to the biologist who wants to study form quantitatively, and computer programs for most of these methods are easily obtained. However, until recently the invariance principle and its implications for the study of biological form have not been carefully considered in the field of morphometrics. We have worked together for over ten years developing a unique approach to the study of biological form. This approach, Euclidean Distance Matrix Analysis, or EDMA (pronounced ed·ma), has considered invariance a central property since its inception (Lele 1991). Real biological questions and the issues that surround them provided the motivation for the development of this approach. We have had the advantage of both statistical and biological perspectives from the beginning of our collaborative work, and believe that this has resulted in a method that is not just statistically correct, but also valuable to biologists. We have struggled with communicating cogent concerns from two different disciplines throughout our collaboration, and have repeatedly stumbled onto new insights that would not have come without scientific questions motivating the development of the methods. Some of these realizations were subtle and complex, others were simpler. Since our collaborative publications have appeared in a number of different journals over many years, we felt it timely to bring our methods, observations, and insights together into a presentation suited to both the biological and statistical community.

Biologists measure forms to obtain quantitative information about specimens. The term measurement connotes quantitative observation, ascertainment of dimensions, and comparison with standards. Referring to an organism as a "large dog" is not particularly informative unless we have an idea of what constitutes "largeness," or have a common frame of reference for dogs. A record of weight and standing height of the "large dog" adds precision to the statement and enables explicit comparison among dogs. Measurements can be used to validate relationships that are qualitatively evident, but can also be used to discover aspects of biological phenomena that may not be discernable from casual observation. Whether metric differences are obvious or obscure, quantitative data can ultimately be used to reveal information pertaining to growth patterns, evolutionary processes, biomechanical design, or phylogenetic relationships. These phenomena are not always directly observable, but may be reflected in morphology.

Most scientific studies of biological morphology focus on one of two primary components: design and diversity (Lauder, 1982). Design refers to the relationship between form and function, a relationship that serves as the basis for several disciplines. Functional anatomy, biomechanics, biochemistry, crystallography, and bioengineering are examples of fields built around ideas pertinent to biological design. The other component, diversity, concerns variation observed among organisms. Variation is observable between and within groups of biological organisms that currently populate this planet, but also across time in the study of evolutionary trends or developmental trajectories. Proper study of design and diversity requires objective criteria upon which to base decisions regarding similarities or differences among organisms. One class of objective criteria is measurement.

1.1 A brief history of morphometrics

The quantitative study of form and form change comprises the field of morphometrics. Morphometrics experienced tremendous activity during the 1980s and 1990s (e.g., Bookstein, 1978; 1982; 1990; 1991; Rohlf and Bookstein, 1990; Reyment, 1991; Marcus, Bello et al., 1993; Rohlf and Marcus, 1993; Marcus, Corti et al., 1996; Small, 1996; Dryden and Mardia, 1998), but its foundations can be traced to the early part of the twentieth century. Sir D'Arcy Thompson is regarded by some as the father of morphometrics, even though a term for this field of study had not been coined at that time. According to Thompson (Thompson, 1992), the purpose of his volume, *On Growth and Form*, was:

> To correlate with mathematical statement and physical law certain of the simpler outward phenomena of organic growth and structure or form, while all the while regarding the fabric of the organism, *ex hypothesi*, as a material and mechanical configuration.

Thompson's work is still regarded as advanced for his time, and his transformational grids are most modern morphometricians' envy. The information communicated by Thompson's transformational grids remains the goal of most morphometric studies (e.g., Bookstein, 1978; Cheverud, Lewis et al., 1983; Bookstein, 1988, 1989; O'Higgins, 1999; Richtsmeier and Lele, 1993).

Two early studies by Franz Boas (1905) and his student Eleanor Phelps (1932) demonstrate that the birth of morphometrics also lies in part in the work of these early anthropologists (Cole, 1996). The reason Boa's contribution is so significant is that he explicitly recognized the arbitrary nature of registration systems used in craniometry (e.g., the Frankfurt Horizontal plane), as well as the implicit assumption that the selected registration points are biologically "stable". Boas suggested that the most favorable superimposition of any two forms would be obtained when all points are considered as having equal weight in the comparison. As pointed out by Cole (1996), Boas' (1905) solution recognized the arbitrary nature of registration systems and anticipated

Sneath's (1967) application of least-squares superimposition by over 60 years.

Julian Huxley also figures prominently in the study of biological shape change because of his generalization of allometry to the study of brain-body relationships (Gould, 1977). Many phenomena of metabolism, biochemistry, morphogenesis, and evolution are governed by the allometric equation in which one characteristic (e.g., length or weight of an organ) can be expressed as a power function of another characteristic (e.g., the length or weight of another organ or of the organism as a whole). Allometric studies usually consist of collapsing measures taken from biological organisms onto two-dimensional, bivariate plots. Most often allometric analysis involves a regression of some biological measure onto another biological variable, the latter being a representative of body size. The projection of measures from individual organisms onto these plots can provide information about the relationship of one organism to another, the relationship of one body part to another part or to the whole, or the change in such relationships over phylogenetic and ontogenetic time.

When used in the study of morphology or morphogenesis, allometry is the study of the relationship between parts, or the proportional relationship of parts to the whole, during either ontogenetic or evolutionary time. In its most general usage, these relationships are often referred to as the relationship between size and shape. Since size and shape are the vocabulary of allometry, allometric studies are closely allied with morphometrics. In allometry, measures are chosen to act as surrogates for body size, and the relationships among these and other variables are interpreted as indicators of shape differences among the organisms. The allometry literature is immense and it is clear that important relationships have been delineated using this approach. Allometry is really a study of proportions, however, and does not carry with it information on the two- or three-dimensional geometry of the organism.

Fred Bookstein's 1978 contribution, *The Measurement of Biological Shape and Shape Change* brought geometry back to the modern study of biological form. Though Bookstein's earliest work (1978) deals solely with two-dimensional data, his intention is to "redefine and reconstruct morphometrics — the measurement of shapes, their variation and change — as a branch of applied modern geometry (ibid: 3)." This definition of morphometrics does not include the term biology, but Bookstein's emphasis is clearly on biological applications. By the early 1980's, morphometrics was redefined as "the empirical fusion of geometry with biology" (Bookstein, 1982). The dual emphasis has remained in theory, but applications have often failed on one or the other of these two foci.

For morphometrics to fulfill its promise of fusing geometry with biology, there must be equal emphasis on the two components. Morphometric techniques need to be designed and applied with biology in mind. Quantitative results must be directly interpretable biologically. In addition, biologists have needs specific to the nature of life. Biologists require methods for correctly analyzing change over time due to growth or evolution, methods that provide for the completion of partial specimens using available data, and methods for predicting, or "retrodicting" in the case of paleontology, the geometry of hypothetical forms based on evolutionary, ecological, functional, and/or developmental considerations. We believe that morphometrics should provide these tools.

Although our focus is on studies in modern biology at the level of the organism, we hope that those seeking tools for use in studies of the cellular or molecular levels of scientific investigation will find value and potential applications in this book. We provide a molecular example through a contribution from our colleague Tim Cole in Chapter 7. We also hope that workers in alternate disciplines such as ecology, geology, paleontology, meteorology, and geography may find value in this book. If our methods are truly useful, they should be applicable to any discipline that requires a statistical tool for the study of geometric relationships when coordinate data are given.

1.2 Foundations for the study of biological forms

We believe, as Bonner and Medawar did, that the success of D'Arcy Thompson's masterful morphometric work, *On Growth and Form*, can be explained by the uniqueness of the author's scholarship and abilities. According to Medawar (1958), D'Arcy Thompson had not only the makings, but the actual accomplishments of three scholars: classicist, mathematician, and naturalist. The fact that these talents were integrated within one person explains his achievement in combining mathematics and biology into a single work that is at the same time a beautiful piece of literature. Neither of the authors of this volume are experts in more than one field. For this reason, we joined forces in authorship. The motivation for comparing biological objects and the knowledge for interpreting those comparisons must come from biology, but the tools for comparison come from statistics and mathematics.

Statisticians must appreciate the nuances of biology and biologists must be cognizant of the assumptions of certain methods and requirements made of the data.

Our collaborations have resulted in the following axioms:

1. **The importance of scientific relevance.** The scientific questions being posed, in our case those pertaining to biology, are of paramount importance. Morphometric methods are developed as tools for answering scientific questions and have little value if the results are not interpretable in terms that relate directly to the scientific question posed. In biology, statistics and mathematics are means to an end. That end is to assist the scientist in the formulation and testing of informed hypotheses.

2. **The importance of biological variability.** Variability can be extreme or subtle, but it is inherent to biological systems. Variability lies at the root of what interests biologists. Variability in biological systems must be modeled as correctly and as accurately as possible. To accomplish this, variability should be modeled stochastically rather than deterministically.

3. **The importance of invariance.** In any geometrically based analysis of biological form, the choice of a coordinate system is arbitrary. Alignment of specimens (registration and superimposition are terms used synonymously with alignment) is required for many morphometric techniques and for all methods of visualization. Alignment is usually accomplished by selecting either a subset of landmarks, a line, or a plane for superimposition of all specimens. When alignment is required, the scientist should be cautioned to run the same analysis trying varying alignments of the same specimens. If results of the comparison of specimens vary between the analyses done using different alignments, then the method used for comparison reflects the effects of the chosen method of alignment rather than the biology of the organisms. The scientific inferences should be invariant to such an arbitrary choice. When scientific inferences are invariant with respect to this arbitrary choice of alignment or coordinate system, they are known in the statistical literature to satisfy the *Invariance Principle* (Berger, 1985). The Invariance Principle will be followed and discussed at length in the context of this book.

These three axioms permeate our book. We lack the contribution of a classicist, but we hope that our collaboration has produced a cohesive blend of mathematics and biology in a readable format that is of scientific value. Our intent is to provide a clear synopsis of Euclidean Distance Matrix Analysis (EDMA), one of several techniques within the class of methods called "geometric morphometrics" (Bookstein, 1978). Our goal is to produce a clear, uncomplicated discussion of EDMA including simple examples that can be joined with a precise presentation of the mathematical and statistical details if the reader so chooses. The discussion will occur within a broader description of alternative geometric morphometric methods in order to clarify certain characteristics of EDMA.

Understanding morphometric techniques can be especially difficult because statisticians and mathematicians often devise the methods, because the methods are communicated in formats that do not always include an explanation of the underlying principles and algorithms, and because biologists, whose familiarity with statistics may only be casual, eventually adopt the methods. We keep the novice morphometrician in mind as we write this volume. We do this by offering explanations of the statistical logic and algorithms at varying levels of difficulty. The first part of Chapters 2 through 4 contains intuitive explanations with uncomplicated arguments of the underlying statistical and mathematical ideas along with analyses of real biological data sets. The second part of these chapters provides a fully rigorous, mathematical treatment of the statistical theory that underlies the methods. The data sets to be used throughout this book are described below.

1.3 Description of the data sets

Three data sets that have been studied in previous publications (Corner and Richtsmeier, 1991; Richtsmeier, Cheverud et al., 1993; Richtsmeier, Valeri et al., 1998; Richtsmeier, Baxter et al., 2000) are used throughout the book to clarify methods and concepts as they are introduced. Any number of examples could be used, but we present data sets with which we have had experience in order to convey statistical and biological observations. Our scientific interest has focused on craniofacial morphology and growth, and our examples reflect that interest. Here, we introduce these data sets in fair biological detail by providing basic descriptions and background relevant to why the data sets are interesting biologically. We provide information pertaining to the collection of these data, but data collection methods are more thoroughly discussed in Chapter 2. Interested readers can go to the original publications for more details pertaining to the data sets and the scientific questions originally addressed through these data. Copies of the original data are available for download from http://faith.med.jhmi.edu or http://www.personal.psu.edu/faculty/j/t/jta10

1.3.1 Data set 1. Mandibular dysmorphology in a genetically engineered animal model.

Trisomy 21 or Down Syndrome (DS) is the most frequent live-born aneuploidy and results in a characteristic spectrum of developmental perturbations affecting many different tissues. Each individual with DS expresses different subsets of the phenotypes that characterize the syndrome. Some of these phenotypes (e.g., heart defects) are less common in DS individuals than other phenotypes. The craniofacial appearance of children with DS occurs in 100% of affected individuals and is immediately recognizable, although variable (e.g., Kisling, 1966; Thelander and Pryor, 1966; Frostad, Cleall et al., 1971; Cronk and Reed, 1981).

   Mouse models provide a powerful approach to the identification of genes whose dosage imbalance contributes to specific aspects of the DS phenotype. Distal mouse Chromosome 16 (MMU16) demonstrates conserved linkage with much of human Chromosome 21 (HSA21), the sequence of which is now known (Hattori et al., 2000). The conserved linkage spans that portion of the chromosome from the most proximal known gene called *STCH* which is located on the q arm of Chr21 to a gene located in the proximal half of the q arm of Chr21 (21q22.3) called *MX1* (Reeves and Irving et al., 1995; Reeves and Rue et al., 1998; Baxter and Moran et al., 2000). The Ts65Dn mouse (Davisson et.al., 1990) has been produced and bred so that it is trisomic for most of this conserved segment of Chr16, meaning that it is at dosage imbalance for many of the genes indicated in DS (Reeves and Irving et al., 1995).

   Ts65Dn mice demonstrate several phenotypic characteristics similar to those of DS. Reeves et al., (1995) demonstrated impaired performance of Ts65Dn mice in a complex learning task indicating that dosage imbalance for a gene(s) in this conserved region contributes to this impairment. Other phenotypes reminiscent of the human DS condition have also been found in the Ts65Dn mouse (Holtzman, 1996; Baxter, 1998). We were interested in studying these mice to see if they

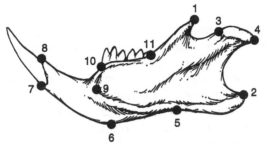

Figure 1.1 Mouse hemi-mandible (left) with 11 left-sided landmarks shown in place. The landmarks are by number: 1, coronoid process; 2, mandibular angle; 3, anterior-most point on mandibular condyle; 4, posterior-most point on mandibular condyle; 5, superior-most point on inferior border of mandibular ramus (joining of angular notch with corpus); 6, inferior-most point on border of ramus inferior to incisor alveolar; 7, inferior-most point on incisor alveolar rim (at bone-tooth junction); 8, superior-most point on incisor alveolar rim (at bone-tooth junction); 9, mandibular foramen; 10, anterior point on molar alveolar rim; 11, intersection of molar alveolar rim and base of coronoid process. Bilateral right-sided points are numbered landmarks 12-22.

expressed patterns of craniofacial dysmorphology. Many genes are conserved across mammals, and the proximate functions of those genes are likely to be conserved as well. We asked whether a conserved phenotypic response could arise from a similar complex genetic insult in humans and mice using the Ts65Dn mouse (Richtsmeier and Baxter et al., 2000). If dysmorphic craniofacial phenotypes are common to Ts65Dn mice, then they can be used to understand the processes by which trisomy for particular genes results in the development of craniofacial dysmorphology.

The Ts65Dn data set consists of landmarks collected from mandibles of aneuploid Ts65Dn mice and euploid littermates. Three-dimensional coordinate locations of 22 mandibular landmarks (Figure 1.1) were recorded using the Reflex microscope (see Chapter 2 for discussion of the instrument for data collection). The analysis of the original data set showed that the crania (including the mandible) of the Ts65Dn mice are significantly different from crania of their normal littermates in ways reminiscent of the difference between the skulls of normal children and those of children with DS (Richtsmeier et al., 1999).

1.3.2 Data set 2: Craniofacial growth in monkeys

The study of growth has long been of interest to biologists. Growth of the skull requires a complex coordination of changes in individual bones as the skull keeps pace with the growing soft tissue organs and spaces that the skull serves to protect and support. Understanding the complex changes that occur during growth of the skull can explain the basis for morphological differences observed in adult forms. Inter specific studies of growth can help us to understand the processes responsible for the production of characteristic features that we use to recognize species differences. Since small changes in growth pattern can be responsible for fairly significant differences in adult form (e.g., Gould, 1977; McKinney and McNamara, 1990; McKinney and McNamara, 1991; McNamara, 1995), studies of growth can provide novel information about phylogenetic relationships among more closely related species.

Growth can be studied using longitudinal data where data points have been collected from a single individual over a period of time. Growth can also be studied using cross sectional data. Cross sectional data consist of a group of measures for each age, but each individual is measured only once so that the sample for an age group does not contain any of the individuals in the previous age group. Both types of data have their limitations and advantages depending upon the research question (Eveleth and Tanner, 1976; Tanner, 1989; Bogin, 1999). We deal exclusively with cross sectional growth data in this volume.

Our cross sectional growth data consist of landmark data collected from skulls of *Macaca fascicularis,* the crab-eating macaque. Male and female skulls from the National Museum of Natural History,

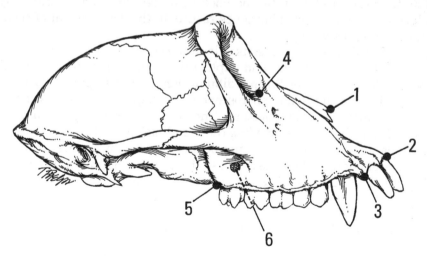

Figure 1.2 Adult *Macaca fascicularis* skull with the location of 6 facial landmarks shown: 1, nasale; 2, intradentale superior; 3, premaxillary-maxillary junction; 4, zygomaxillare superior; 5, maxillary tuberosity; 6, posterior nasal spine (located on the sagittal plane).

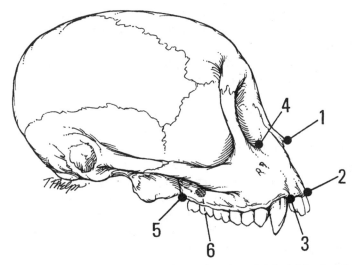

Figure 1.3 Adult *Cebus apella* skull with the location of 6 facial landmarks shown. These landmarks are homologous to those plotted on the adult Macaca fascicularis skull in Figure 1.2. Landmarks shown include: 1, nasale; 2, intradentale superior; 3, premaxillary-maxillary junction; 4, zygomaxillare superior; 5, maxillary tuberosity; 6, posterior nasal spine (located on the sagittal plane).

Smithsonian Institution, were aged according to dental eruption patterns (Richtsmeier and Cheverud et al., 1993). Landmarks were digitized directly from the external surface of the face, neurocranium, and cranial base of these skulls using the 3Space digitizer (see Chapter 2). Landmarks used in this book represent a subset of the available facial landmarks (Figure 1.2).

In addition to the landmark data collected from a large collection of skulls of *Macaca fascicularis,* we have a comparable data set collected from skulls of *Cebus apella*, the capuchin monkey (Corner and Richtsmeier, 1991) (Figure 1.3). These species are members of differing infraorders of the suborder *Anthropoidea*. The split between the New World monkeys (Infraorder: platyrrhini, of which *Cebus apella* is a member) and the Old World monkeys (Infraorder: catarrhini, of which *Macaca fascicularis* is a member) dates to a time before the Oligocene. Beyond the geographic separation that has been maintained over millions of years, New World monkeys differ from Old World monkeys in many respects including, but not limited to social structure, overall body size, and craniofacial features.

These data sets enable us to look specifically at differences in growth patterns between two species whose phylogenetic relationship is known. A combination of tools based on EDMA will allow us to determine differences in immature morphologies, differences in growth patterns, and differences in adult morphologies, providing an understanding of the role of growth pattern in the determination of adult differences in cranial morphology.

We will also use these data sets to demonstrate the use of EDMA tools in experiments in evolutionary morphology. Beginning with an immature individual of either species, we will demonstrate how our methods can be used to apply an observed growth pattern to that form, producing a hypothetical morphology. The hypothetical morphology represents what an immature form would look like if it had followed the growth trajectory implemented in the experiment. We believe that these tools may prove powerful in future analyses of phylogenetic relationships, evolutionary trajectories, and developmental or phylogenetic constraints.

1.3.3 Data set 3: Human craniofacial dysmorphology and growth

This data set consists of landmark coordinate data collected from computed tomography (CT) scans of children diagnosed with isolated sagittal synostosis (Richtsmeier et al., 1998). Sagittal synostosis is the pre-

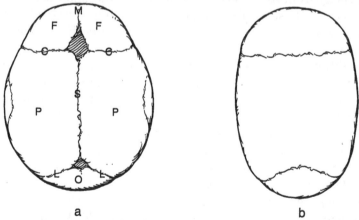

a   b

Figure 1.4a The developing human skull showing isolated bones (frontal (F), parietal (P), occipital (O)) and sutures that lie between the bones (coronal (C), sagittal (S), lambdoid (L)). **1.4b** The typical dolichocephalic shape of a skull when the sagittal suture closes prematurely (sagittal synostosis).

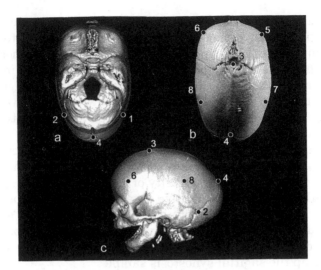

Figure 1.5 Three-dimensional reconstruction of a computed tomography (CT) scan of a child with premature closure of the sagittal suture. Figure 1.5a shows a superior view of the skull with superior surface of neurocranium removed. Figure 1.5b shows a superior view of the skull with external surface of the neurocranium intact and demonstrating approximate placement of the following landmarks: 3=bregma, 4=lambda, 5= right frontal boss, 6= left frontal boss, 7= right parietal boss, 8=left parietal boss. Figure 1.5c provides a lateral view of the left side of the skull with the following landmarks indicated: 2= left asterion, 3=bregma, 4=lambda, 6=left frontal boss, 8=left parietal boss.

mature closure of the sagittal suture. Isolated, or nonsyndromic synostosis of the sagittal suture is fairly common (3 to 5 per 10,000 births), occurring more frequently in males than in females (Cohen, 1986). The developing neurocranium is made up of a number of roughly shell shaped bony plates that align with one another at joints or articulations called sutures. The sagittal suture lies between the paired parietal bones (Figure 1.4.a). The anterior-most point of this suture lies at the anterior fontanelle and in the more mature individual is defined by the landmark bregma, which marks the intersection of the paired frontal and paired parietal bones. The sagittal suture runs from the anterior fontanelle (or the landmark bregma) between the two parietal bones along the top of the skull until it intersects with the right and left segments of the lambdoid sutures that separate the parietal bones from the occipital bone. The intersection of the sagittal suture with the lambdoid suture is marked by the landmark lambda. In most children, the sagittal suture remains open until adulthood when it begins to ossify. If the suture is closed prematurely, the neurocranium takes on a characteristic shape that is long and narrow relative to a normal skull (Figure 1.4b). This condition is referred to as dolichocephaly or scaphocephaly. There is the potential problem of increased intracranial pressure and eventual neurological damage in children diagnosed with sagittal synostosis. For this reason children with a positive diagnosis of sagittal synostosis almost always undergo surgical suture release and in some cases, reconstructive surgery.

Landmark data were collected from CT scans of patients ranging in age from six weeks to seven years with a positive diagnosis of sagittal synostosis (Figure 1.5). The majority of the patients were aged from 4 to 14 months. Normal, comparative data were collected from CT scans of the Bosma collection (Shapiro and Richtsmeier, 1997), a sample of normal pediatric skulls that were aged by tooth eruption patterns and chosen to correspond as closely as possible in age to the individuals that make up the sagittal synostosis sample.

CHAPTER 2

Morphometric Data

"We may want now and then to make use of scanty data, and find a rough estimate better than none."

D'Arcy Thompson (1992)

Data are collected from any object of interest in order to capture salient features of overall form and characteristics of specific regions defined by the scientific problem under study. The purpose of the study might be descriptive, exploratory, comparative, or explanatory, and the type of data used may vary with the purpose of the analysis. In Part 1 of this chapter, we introduce a variety of morphometric data types, but we emphasize landmark data. Because we are primarily concerned with biological inquiry, we explore the biological meaning of landmarks, then discuss various methods of landmark data collection, including some pitfalls, and provide a design for measurement error studies. In Part 2 of this chapter, we give a brief introduction to matrix algebra to clarify the mathematics necessary to work with landmark data. These concepts are then used to develop generalized approaches to the study of measurement error for landmark data.

2.1 Types of morphometric data

2.1.1 Unidimensional measures

A single uni-dimensional measure may be all that is needed to adequately describe the form under study for a particular research purpose (Figure 2.1.a). Measures like crown-rump length, head circumference, and wing span give a general indication of the overall size of a biological form and may provide enough information to determine

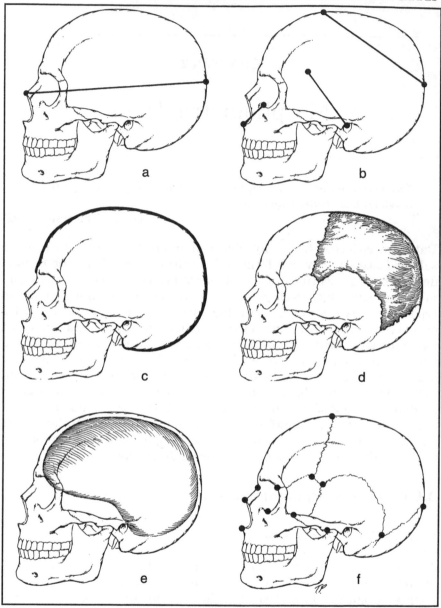

Figure 2.1 Aspects of a human skull used to demonstrate various data types used in morphometric analysis. a) a single unidimensional measure (linear distance from nasale to lambda) that could represent overall length or size of the skull; b) a set of unidimensional measures made up of three linear distances with no common endpoints that measure different aspect of skull form; c) the outline of the neurocranium as recorded from a lateral view. Here we show an open curve, but closed curves can be used; d) the surface of the parietal bone; e) size of the interior of the skull measured as a volume; f) the form of the skull measured as a set of landmark locations in three dimensional space.

if an object lies within the predetermined size distribution of a population. For example, the standard growth chart for infants plots the mean and standard deviation for body length, body weight, and head circumference against age. New data points can be plotted and evaluated in terms of the population norms for these measures.

2.1.2 Groups of linear distances

Sets of linear distances can be designed to summarize a form (Figure 2.1.b). Linear distance data are collected using measuring tapes or sticks, hand-held calipers or other anthropometric devices, or computed from the coordinates of landmarks. Univariate statistics enable the study of each of the measures by itself. However, scientists often want to study the relationships among these various measures. Two linear measures can be combined as a ratio to provide an index of the relationship between these measures within a form (e.g., wing length and breadth). Such indices are often referred to as measures of shape and are commonly used in biology. (To learn more about the use of ratios in biology see Atchley, Gaskins et al., 1976; Atchley, 1978; Mosimann and James, 1979).

Multivariate statistical techniques enable the study of multiple measurements by considering the relationships among them. The usefulness of these methods in the analysis of biological organisms was quickly recognized and multivariate techniques gained popularity among biologists by the late 1960s. The use of multivariate statistics did more than offer a way to look at a combination of linear distances. These methods also offered a way to simultaneously consider multiple measurements representing variables of different types (e.g., morphological, ecological, nutritional, or life history measurements). There are many excellent statistical texts that treat multivariate techniques (e.g., Mardia, Kent et al., 1979). Kowalski (1972) provides a clear summary of multivariate techniques, as well as an informed cautionary note regarding the potential scientific (biological) ambiguity of results and the impact of this on interpretation and communication of results.

2.1.3 Outlines, surfaces and volumes

Outline data are two-dimensional representations of the boundary of a form (Figure 2.1.c). Examples of such data include the outlines of a skull, a joint articulation, an insect wing, or a granule of grain. Techniques for quantitative analysis of such data are most commonly based on either Fourier analysis or on Eigenshape analysis (e.g., Ehrlich and Weinberg, 1970; Lestrel, 1982; Ehrlich, Pharr et al., 1983; Lohmann, 1983; Ferson, Rohlf et al., 1984; Lestrel, 1989; Lohman and Schweitzer, 1990; MacLeod and Rose, 1993; Rao, 2000).

One major drawback of outline data is that one cannot unambiguously establish correspondence between the outlines of two objects. A one-to-one correspondence between data sets collected from various objects is critical to meaningful comparisons. The lack of correspondence between outlines makes it unlikely that the analysis of outlines will enable the identification of localities that are different between forms (Cole and Wall, 2000). On the other hand, outlines can provide certain information about the form (e.g., curvature) that cannot be captured by landmarks (Bookstein et al., 1982; Ehrlich et al., 1983; Read and Lestrel, 1986; see Figure 2.2).

A surface is the area of the outer or inner face of an object (Figure 2.1.d). A surface can be recorded as an array of points in three-dimensional space. A volume can be thought of as the amount of space within a closed surface measured as a single number in cubic units (Figure 2.1.e). An alternate measure of volume is the three-dimensional coordinates of points that map the contents of that closed surface. Data arrays representing surfaces and volumes are large, and statistical tools for summarizing and comparing the topography or shapes described by these data sets are not well developed.

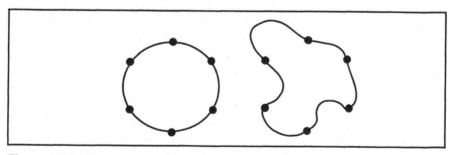

Figure 2.2. Outlines can provide certain information about the form (e.g., curvature) that cannot be captured by landmarks. In this figure identical sets of landmark data are recorded on two forms. In the example shown here, comparison of the sets of landmark data collected from the two forms would determine that the two forms are the same while outline information shows clear differences between the two forms (after Read and Lestrel (1986), figure 2).

2.1.4 Landmarks

A landmark is a point in two- or three-dimensional space that corresponds to the position of a particular feature on an object of interest. For example, in the study of osteological remains, a landmark might be defined as the point that marks the scar of a muscle insertion on a bone, the intersection of two or more bones at a cranial suture, or the foramen that marks the path of a neurovascular bundle. We choose to focus on landmark data because we want to analyze data that have unambiguous correspondence between the forms being compared. Issues relating to correspondence are discussed below. The mathematical reasons for preferring landmark data are developed later in this book (see especially Part 2 of Chapters 3 and 4).

One of the first decisions that a researcher must make is the selection of landmarks for approaching a research question. This choice is necessary but subjective. Landmarks should be chosen in order to provide an adequate representation of the object under study, but the research question will also dictate the landmarks chosen for study. There are no concrete rules for making this choice. Our rules of thumb include: 1) know your specimens, 2) make a list of all potential landmarks, 3) collect data from as many landmarks as are feasible. Once data are collected and analyses are underway, the number of landmarks can be reduced on the basis of measurement error studies, or redundancy of information from landmarks.

Landmarks collected from biological objects may be classified into three general groups discussed below. Similar groups can probably be formed from landmarks taken from nonbiological objects but our lack of experience with non-biological data sets discourages us from generalizing.

a) Traditional landmarks

Traditional landmarks are precisely delineated points corresponding to the location of features of some biological significance. There are two classes of traditional landmarks: those whose definition is not dependent upon a coordinate system, and those whose definition is tied to a particular orientation or a coordinate system. Examples of the first class include landmarks such as nasion and nasale. Nasale is defined as the distal end of the internasal suture, while anterior nasal spine is defined as the intersection of the right and left nasal spines of the maxilla (Figure 2.3). Both of these points can be located on a skull regardless of the orientation of the skull or its position in the laboratory. The exact

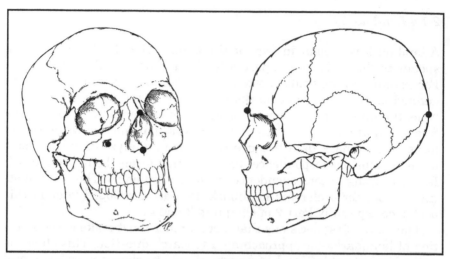

Figure 2.3. Examples of two classes of traditional landmarks. Shown at left are the landmarks nasale, anterior nasal spine and maxillary foramen. The location of these landmarks can be precisely given without reference to a coordinate system. Shown at right are the landmarks glabella and inion. These landmarks can only be found when the skull is placed in the Frankfort horizontal orientation and the skull is viewed from the side. Thus, the placement of these landmarks requires a specific coordinate system.

coordinates of either of these points change according to the position of the skull, but their location relative to each other and to other landmarks on the skull remains the same regardless of position or orientation of the skull.

  The second class of traditional landmarks consists of precisely delineated points corresponding to the location of features of some biological significance but whose definition is dependent upon a particular coordinate system. Examples include glabella and inion (Figure 2.3). Glabella is defined as "the most forward projecting point of the forehead in the midline at the level of the supra-orbital ridges above the naso-frontal suture" (Bass, 1971). The definition of this landmark requires that the skull be positioned in a particular way in order for that landmark to be located. This means that the location of this landmark may change depending upon the orientation of the skull. Even when orientation is explicitly defined (e.g., the Frankfort horizontal plane), the exact orientation may never be duplicated when collecting data from other subjects, resulting in an additional source of potential error. For this reason, landmarks that require orienting of the specimens for data collection are less desirable.

b) Fuzzy landmarks:

A fuzzy landmark is a point corresponding to a biological structure that is precisely delineated and that corresponds to a locus of some biological significance, but that occupies an area that is larger than a single point in the observer's reference system (Valeri et al., 1998). The definition of a fuzzy landmark usually includes a positional reference (e.g., centroid, apex) that corresponds with a place on the feature that best represents it as a point. Because fuzzy landmarks must be located on a larger structure or surface, there is the possibility that more measurement error will be included in their location (see Valeri et al., 1998). Fuzzy landmarks become necessary when portions of the form under study are made up of relatively large, smooth surfaces or features and do not include sufficient traditional landmarks. Biological examples of such features include the squamous portion of mammalian cranial bones, the articular surface of the astragulus, the articular surface of the capitulum of the humerus, the basin between two shearing facets on a tooth, the outer capsule of pome fruits, the external surface of some crab carapaces, the leading edge of a canine tooth, and the ischial tuberosity. The example in Figure 2.4 shows the left frontal

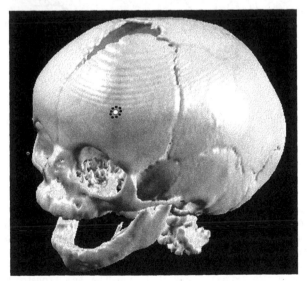

Figure 2.4 Three-dimensional reconstruction of a computed tomography scan of an infant showing an example of a fuzzy landmark. The frontal boss is a swelling on each frontal bone. It is a rather large area. The black points shown in the figure depict the placement of a single point to localize this feature in a series of data collection trials. The white point provides an estimate of the average location of these trials.

boss, a large eminence on the left frontal bone of human juveniles, the center of which can be considered a "fuzzy" landmark. Due to their nature, placement of fuzzy landmarks on an object may require multiple data collection episodes with the average of the coordinates from several data collection episodes being used for analysis.

c) Constructed landmarks:

Constructed landmarks are points corresponding to locations that are defined using a combination of traditional landmarks and geometric information. Oftentimes, there exist surfaces of interest that are void of traditional landmarks and fuzzy landmarks. In this instance, land-

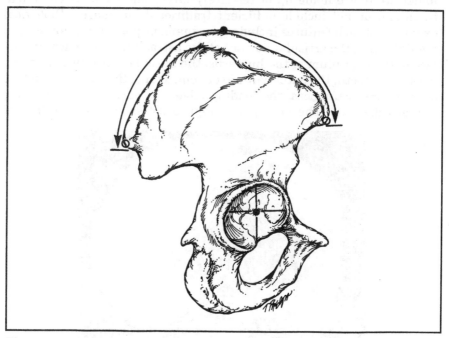

Figure 2.5. Two constructed landmarks shown on a human hip bone. The first is constructed by locating two traditional landmarks, the anterior superior iliac spine and the posterior superior iliac spine (marked by open circles) and connecting these landmarks by a line that follows the arc of the iliac crest. The midpoint of this arc (shown as a filled circle) is recorded as a constructed point. The second constructed point is located by drawing two lines at right angles to each other across the acetabulum (socket for the ball of the femur), one line measuring the width of the acetabulum, the other measuring the height. The intersection of these lines cross on the acetabular surface is recorded as a constructed point.

marks can be constructed. Let us say that data are being collected from the pelvis. Several features of the pelvis can serve as traditional landmarks (e.g., McHenry and Corrunccini, 1978; Li and Richtsmeier, 1996), but the large alar surfaces are void of landmarks. Figure 2.5 shows several landmarks located on a human innominate (hip) bone. A landmark can be constructed by using the location of two traditional landmarks, say anterior superior iliac spine and posterior superior iliac spine (marked by open circles on either side of the bone in Figure 2.5) and connecting these points by an arc that stretches along the surface of the iliac crest. A single midpoint along this arc (as shown in Figure 2.5) or several points at specific intervals along the arc can be identified and their location recorded as a constructed landmark. The same could be done using a combination of any of the traditional landmarks defined on a form. A string that hugs a surface can be useful in defining constructed points. The "string" can be an actual one, hand-held by the researcher, or can take the form of an algorithm that computes the shortest distance between the two traditional landmarks along a given surface captured in digital format.

Constructed landmarks are those that might easily be placed by computer-based algorithms for landmark locating. Although computer automation of data collection would significantly reduce the amount of time required for data collection, caution must be taken. Biological variability can figure into landmark location in unexpected ways. For example, if a point were defined as lying midway along the shortest possible distance between two points, it is possible that the shortest distance between two points may not cross the surface of interest, but instead pass through or wrap around an alternate surface. When total automation of landmark data collection by computer algorithm is a viable option, continued supervision by the researcher is critical in correctly locating all landmarks.

Whether located by a human or a computer, the biological and statistical properties of constructed landmarks are different than those of traditional landmarks, and this needs to be considered during data collection, analysis, and interpretation. As suggested above, a lack of correspondence between constructed landmarks collected on different forms may occur. The locus midway between two landmarks along a surface on one object may not correspond biologically with a locus midway between the same two points along the homologous surface of another object. Error in the location of constructed landmarks is dependent upon error in the location of each of the traditional landmarks and error in the placement of the constructed landmark.

Population variability in constructed landmarks will reflect population variability in the traditional landmarks and in the surface used in their construction.

2.2 Landmark homology and correspondence

In many biological studies that use landmark data, researchers refer to landmarks as *homologous*. A homologue, as defined by Richard Owen (1848), is "same organ in different animals under every variety of form and function." Though the definition seems simple enough, the concept of homology continues to stimulate debate (Hall, 1994; 1999). Some of this discussion stems from common usage of the term homology as the "correspondence of parts" with no specification about whether the parts correspond with respect to structure, phylogeny, development, or on the basis of another relationship. Biologists (particularly paleontologists, anthropologists, and natural historians) define homology as the establishment and conservation of individualized structural units in organismal evolution (Müller and Wagner, 1996), or more simply, similarity in structure due to descent from a common ancestor. When used in mathematics and engineering, homology does not imply phylogenetic relationships. But even within a discipline, the definition of homology is applied inconsistently, and a practical definition remains elusive (Wagner, 1989; Roth, 1988).

Although we cannot cover the topic adequately here, the reader needs to be aware of the nuances associated with the concept of homology. For example, characters can appear to be homologous (similar due to common descent) but actually represent homoplasy. Homoplasy exists when a character is present in two species but is not possessed by all intervening ancestors or by the most recent common ancestor of the two or more species under consideration. A character can evolve more than once in unrelated species due to similar selection pressures acting on those species, as in the case of convergent evolution (e.g., wings evolved separately among the mammalia and among the birds; leglessness evolved separately in snakes and in certain lizards).

This situation forces careful use of the term homology when selecting features for morphometric comparison. In this monograph, our examples investigate differences in form within single species or within very closely related species. However, care must be taken in the use of the term *homologous* when landmarks designate features on organisms of diverse phylogenetic history. We believe that the term homologous should be reserved for the analysis of features that are shared due to common descent (see below). We suggest the term *correspondence* be substituted for all other relationships so as not to confuse phylogeny with other bases for comparison. Phylogenetic history is known for relatively few organisms, and other important issues need to be evaluated (see below). We offer the following outline of useful terms as a tentative nomenclature and a basis for further discussion. The proposed categories of landmark correspondence are not mutually exclusive.

a) Homologous landmarks

Homologous landmarks are those landmarks chosen to represent features on organisms that are similar due to a phylogenetic relationship. The organisms being compared thus share a common ancestor, and the feature under study is present in all organisms under consideration due to each inheriting it from the common ancestor. This meaning refers to the phylogenetic, evolutionary, or historical definition of homology (Wagner, 1989). If the organisms being compared represent members of a single species, then corresponding landmarks are homologous by definition because all members of a species descended from a common ancestor. But if organisms from different species are being compared, the landmarks used can only be considered homologous if the phylogenetic history of the organisms is known. Lack of this type of information forces the researcher to consider alternate terms when referring to landmarks (see Wagner, 1989; VanValen, 1982; Muller and Wagner, 1996).

b) Structurally corresponding landmarks

Structurally corresponding landmarks are those landmarks chosen to represent structurally similar features on organisms. Structural correspondence of parts is relatively easy to establish morphologically and is a more conservative definition of relationships, especially when the phylogenetic history of the organisms under consideration is not known. An example would be a comparison of the femur of a dog and the femur of a dinosaur; two organisms separated greatly in terms of evolutionary time and phylogeny, but with limbs that show basic structural correspondence. Femora of these two taxa share many distinguishing features, and comparable landmark data can be collected from the two species for that reason. Landmarks identified on structurally corresponding parts may display functional correspondence, as in the case of bat and bird wings, or they may not (e.g., dorsal fin of a dolphin and hind leg of a salamander).

c) Functionally corresponding landmarks

Functionally corresponding landmarks are those landmarks that are defined on features that have similar function in the organisms under consideration. Insect wings and bird wings are textbook examples of structures with functional correspondence, but differing phylogenetic histories (i.e., insect wings are functionally similar to bird wings but not homologous). Identification of functionally corresponding features usually requires in-depth biomechanical analysis to confirm functional correspondence.

d) Developmentally corresponding landmarks

Developmentally corresponding landmarks are chosen to represent features that are the developmental precursors of specific adult structures. Developmentally corresponding landmarks may be similar in structure between organisms, or they may be very different. This is because the structures of different organisms can be homologous even though their developmental pathways have come to differ substantially due to evolutionary change of developmental pathways (Müller and Wagner, 1996; Fuytuma, 1995; Rice, 1998; Hall, 1999). When form is more conservative than the developmental route by which it is achieved, the form of an organism or parts of that form retain an identity through evolutionary time, but the molecular blueprint or developmental pathway of the structure may change (Futuyma, 1995; Raff, 1996; Rice, 1998). For example, Jacobson and Sater (1988) have shown that the optic cup is necessary to lens formation in some species of vertebrates, but not in others. Non-corresponding developmental processes can therefore produce corresponding structures. Hall (1999) provides an excellent discussion and additional examples of the variability of developmental processes that produce homologues. The need for consideration of this special case of correspondence may be rare in the analysis of landmark data, depending upon whether an ontogenetic stage (embryological vs. postnatal), the adult form, or the developmental pathway is the focus of the research. Molecular genetics and evolutionary developmental biology have shown us that conserved mechanisms of development may indicate homology of morphologically divergent characters, that non-homologous structures can develop from homologous anlagen, and that the molecular mechanisms of development may be put to use in varying contexts producing novel morphologies (Müller and Wagner, 1996). Resolution of these possibilities in any particular case requires careful evaluation of molecular mechanisms, developmental programs, and morphological similarity.

It is obvious from the above classification that there are landmarks that could be shared among organisms on more than one of the biological bases outlined above. The particular orientation of the research and the research question that is being posed will dictate the actual choice of landmarks. The choice of which specific landmarks should be used in a study is arbitrary but should be made on the basis of the scientific question being posed.

2.3 Collection of landmark coordinates

Landmark coordinates are the location of points in space recorded along X- and Y-axes in two-dimensions and along X-, Y-, and Z- axes in three-dimensions. Landmark coordinate data can be collected directly from whole specimens, from serially sectioned specimens, or from images of specimens. The following is not meant to be a comprehensive review of available techniques and equipment, but rather a listing and description of some of the types of devices already used in biological research and available to the scientific community. We mention by name only those instruments with which we have had personal experience or with which we have had personal contact with users. Parts of this discussion follow a paper prepared by Luci Kohn (1994), presented at the Morphometrics workshop held during the International Conference of Vertebrate Morphology, Chicago, 1994.

2.3.1 Commonly used devices for collection of landmark coordinate data from whole specimens

Diagraph

The diagraph is a machine that allows collection of landmark coordinate data using only paper and pencil. A stable vertical stand anchors necessary parts of this device. The object to be digitized is immobilized on a stage secured to the stand over a piece of ordinary paper. The paper is taped to the writing surface so that it does not move during data collection. Two pointers are also mounted on the stand. These pointers are plumb with each other and move in concert in X and Y space. One of the pointers is used to locate the landmark on the object being digitized, while the other pointer traces this location in X,Y space near the bottom of the stand. When a landmark is located by touching a location using the upper pointer, the lower pointer is depressed to mark the landmark location in X,Y space on the paper. That location is then labeled. The Z coordinate is read from a scale mounted on the stand that indicates the height of the upper pointer that is in contact with the landmark and recorded. This is done for each landmark. After the X and Y coordinates of all landmarks are marked on the paper, they are most easily entered into a computer using a two-dimensional digitizer pad. The Z coordinate associated with each X,Y pair is entered by hand. If a digitizing pad is not available, a coordinate system superimposed onto the paper could allow the researcher to determine the X and Y coordinates. These could then be entered into a computer along with the Z coordinate. Diagraphs can be custom built or bought from anthropometric instrumentation supply houses for less than $1500. The diagraph is highly portable and does not require a computer during the data collection phase. It is technically simple and easy to use.

Digitizers

A digitizer is a machine that enables either two- or three-dimensional coordinate data to be collected directly from a specimen by direct contact of a mouse (2D) or stylus (3D) to the point of interest. Two-dimensional tablets are inexpensive, manufactured by many companies, and come in a large selection of designs and sizes. All two-dimensional digitizers are intended to record the location of specified points of interest within a 2D grid. Three-dimensional digitizers are more expensive, more limited in availability and design, and operate using either sonic or electrostatic signals. The 3D digitizers are able to sense the location of a pointer or indicator in a defined 3D volume and to assign a specific coordinate location to that point within the defined volume.

The Polhemus 3Space tabletop digitizer is the machine with which we have had the most experience (see Corner, Lele et al., 1992), but it is no longer available commercially. The next generation of 3D digitizers available from Polhemus includes the 3DRAW digitizing tablet. The 3DRAW enables direct digitization of specimens in either two- or three-dimensions. The 3DRAW includes a stylus that is shaped like a pencil connected to a tablet. The 3DRAW uses electromagnetic technology that precisely measures the location of the stylus tip and, if desired, its orientation in reference to the source located within the tablet. 3DRAW requires connection to a host computer via the RS-232C serial port. Data can be continuously input as the stylus moves through space around the object, or single data points can be collected. A data capture switch is located on the stylus as well as on a footplate. We have found the capture switch on the stylus to be less useful for landmark data collection, as depressing the switch invariably causes movement of the stylus, which should remain motionless and in contact with the landmark. The footplate is relatively easy to use, especially if placed on a tabletop and depressed by hand. Once the data capture switch is depressed, the X, Y, and Z coordinates of that point are written to an ASCII file on the host computer. The 3DRAW tablet is portable (approximate dimensions: 21.5" x 16.4" x 3.1"), weighing 5.5 pounds. Objects are measured within a volume measuring 17.75" x 12.75" x 12" which limits the user to smaller specimens. The resolution of the 3DRAW is quoted by Polhemus to be within 0.005, with an accuracy of 0.01 inch root mean square. The cost of the 3DRAW exceeds $5000. It is portable, but requires a computer. Elementary software for digitizing landmarks is included.

Immersion Corporation produces the MicroScribe-3D, a highly portable digitizer with a mechanical arm and sensors that track the position and orientation of the stylus tip. Digitizing software available with the MicroScribe-3D allows graphical modeling using points, lines, polygons, or splines. Data can be exported in several standard formats, and drivers are available that allow the digitizer to function with graphic modeling packages as well as spreadsheet software. The digitizer requires a computer connection but is compatible with PC, Macintosh, and Silicon Graphics workstations. Three models are available which differ in cost and resolution. Steve Leigh (personal communication) has reported an average error estimate of .5 mm (calculated as the average distance between multiple digitizations of a single landmark). He reports a lack of statistically significant differences between linear distances calculated between points digitized using the MicroScribe-3D and those same linear distances measured using calipers. We have not conducted measurement error studies using this digitizer.

Optical metric systems

The Reflex 3-Metric System includes instruments that provide measurements from stationary objects in three dimensions without directly touching the specimen and requires interface with a personal computer. MacLarnon (1989) described these instruments in detail. The Reflex Microscope involves the use of the stereoscopic ability of the observer to pinpoint the surface of an image to within a few microns along the Z-axis, from a distance of typically 65mm. This allows the examination of objects that have a very uneven surface. An illuminated spot of a selectable size is used to precisely locate the feature to be measured. The selected size of the illuminated spot allows very precise delineation of the part of the image under examination. The act of measurement does not degrade the image and no direct contact is made with the specimen under view. The microscope represents a hybrid between a small coordinate measuring machine and a surface profiler and is ideal for measuring three dimensional distances and angles between small surface features with irregular surfaces. It is also capable, in principle, of quantifying sub-surface features in transparent or semi-transparent materials, although to do this, the influence of the material refractive index has to be accounted for.

When digitizing, the object is viewed through an adapted stereoscopic microscope where a small light spot appears in the field of view. Using the motorized stage the light spot can be moved until the location of the relevant surface feature coincides with the spot. The depth coordinate is set by the observer using stereoscopic vision to locate the spot "on" the surface of the object. The X-, Y-, and Z-coordinates are monitored continuously via linear encoders and can be stored on command in the computer so that single points or streams of points that describe curves can be collected and written to a file for future analysis. The object stage, driven by motors, can be guided manually using high resolution joysticks or automatically under software command, allowing movements to be constrained to predetermined planes or directions if required or to pre-set patterns.

Measurement error studies of earlier models of the Reflex instruments are available (Scott, 1981; Setchell, 1984; Speculand et al., 1988a, 1988b; see also MacLarnon, 1989). The latest version of the microscope (January, 1999) provides an absolute accuracy of better than $\pm$ $3\mu m$ in XY and $\pm$ 4 μm along the Z-axis. Repeatability is said to depend largely on the skill of the operator and is expected to be better than ± 2 *microns* in X- and Y-, and ± 4-*microns* in Z-, using the highest standard magnification. Reflex Measurement Ltd. has developed an auto-focusing facility device that automates measurement along the Z-axis. We have not yet published our repeatability studies using this device.

Surface scanners

Surface scanners are devices that make use of a variety of digital 3D sensing techniques to produce a high resolution image of the surface of objects. The surface is recorded as a series of closely spaced points in 3D space. For example, Coward and McConathy's (1995) facial images each consist of 256,000 points. Landmarks can be identified on imaged surfaces, and coordinate data can be collected from those surfaces. A large number of surface scanners are available for use in the study of biological morphology. If a laser scanner is readily available, this is a reliable way to collect landmark data (Kohn and Cheverud, 1992). Moreover, if specimens are of a difficult size, scanning an entire specimen enables one to either scale the specimen up (as in the case of Eocene primates) or down (as in the case of Neoceratopsian dinosaurs) for digitization. This enables data collection from an organism that would be difficult or impossible to digitize in its natural state.

2.3.2 Digitizing whole specimens in multiple phases

When collecting coordinate data from whole specimens, it is essential that the specimen remain stable. If the specimen is jostled or moved during the digitization process, the integrity of the relative location of landmarks is lost and the digitization procedure needs to be restarted. To ensure stability, it is necessary to secure the specimen in place. Pressing a specimen into a mound of clay is a common practice. When this is done, a portion of the specimen is hidden from view while the remainder is being digitized. After digitizing one side of the specimen, it must be repositioned in order to collect landmark data from the other side, yielding two (or more) subsets of data for each specimen. In order to join the two landmark subsets into a common coordinate system, it is necessary that a specific group of landmarks, visible from both views, be collected from the specimen in both orientations along with the set that can only be seen from one side. Once the data collection is complete, two subsets of data exist for every specimen, and each subset includes a group ($N \geq 3$) of shared or *common* points that are recorded from both views. The two subsets are joined into a single, three-dimensional configuration by simply translating and rotating the common landmarks is recorded in one view (e.g., landmarks from the right side of an object) to match exactly the coordinates of the same common points collected from the other view (e.g., left side). Computer programs can facilitate this process. Since the relative location of the common landmarks are identical in the first and second set (barring significant measurement error), the coordinates of the common landmarks can be matched exactly, and the relative location of the remaining landmarks collected from the two sides are brought into this single coordinate system.

2.3.3 Landmarks from serial sections or captured images

Embryonic, fetal, and other histological and anatomical specimens are often serially sectioned in order to reveal aspects of internal structure. Historically, reconstructions of these sections have been done to reveal the three-dimensional structure of internal features. Two-dimensional data can be collected from each slice, and when special precautions are taken to enable registration of the sections to one another, a surrogate coordinate of the third dimension can be added using information from the relative location and thickness of the particular slice. Working with serially sectioned specimens introduces new avenues for potential error. First, error can be introduced by the methods of registration and alignment used in reconstruction of the sectioned specimen. Additionally, Lozanoff (1992) points out that tissues may distort or shrink if subjected to fixation and dehydration. Distorted images may result even when tissues are normal if the optical properties of the video camera or the configuration of the video board are flawed (Lozanoff, 1992). These problems may seem overwhelming to the untrained histomorphologist who wants to obtain reliable landmark data from histological sections. Lozanoff (1992) has demonstrated an alternative approach that involves the recording of images of sectioned surfaces prior to microtome sectioning. He provides the accuracy and precision of measurements taken from computerized three-dimensional reconstructed models made from these images.

2.3.4 Landmarks from images

The collection of two-dimensional landmarks from projections of forms onto film or paper has a long and rich history in biological research. Camera Lucida drawings of objects viewed under a microscope, clinical x-rays taken using a roentgenographic cephalometer, and simple photographs taken with attention paid to scale can all be used as a source from which to collect landmarks. All that is required is that a grid of known scale be superimposed over the stationary image, and points located on the biological form and their coordinates be noted for later analysis. Although collection of these coordinates can be done with pencil and paper, two-dimensional digitizing tablets that attach to a computer increase the speed and precision of data collection. These digitizing tablets are inexpensive and ubiquitous (see Digitizers, above).

Stereophotogrammetry has been used to collect three-dimensional coordinate data in biological analysis (e.g., Creel, 1978; Hartman, 1986). This method requires a specialized camera system. Two exposures, a stereo pair, are made of a single object. The pair represents the object from two slightly different perspectives. A specialized stereoplotter, of the sort used in cartography to construct contour maps from aerial photos, is then used to project a dot onto a landmark location. The position of the dot in X, Y, and Z space is measured by the stereophotogrammetric system. The usual stereophotogrammetric systems (including cameras and plotters) are expensive. Simplified short base stereophotogrammetry instruments were developed for clinical applications in the 1960s (Beard, 1967; Burke and Beard, 1967) and have been used to measure change in facial morphology due to growth. Biostereometrics is an expansive field with a rich history and a vast literature, which we will not attempt to document here.

Diagnostic medical imaging in the form of computed tomography (CT) and magnetic resonance imaging (MRI) has become routine in medical practice and fairly common in biological research. The availability of large data sets at little or no cost and public domain imaging software for viewing images makes this type of data a practical research alternative for most laboratories. For example, anatomical data are available via the Internet from the Visible Human Project (http://www.uke.uni-hamburg.de/institute/imdm/idv/forschung/vhp/index.en.html) and the Visible Embryo project. Software tools that are available free of charge via the Internet include NIH image, available via the Research Services branch of the National Institute of Mental

Health (http://rsb.info.nih.gov), and *etdips* (http://www.cc.nih.gov/cip/software/etdips/). There are many, many software packages that can be purchased for use in analysis of medical images.

We focus exclusively on landmark data collection from CT images in this chapter, since these are the data that we find most useful and accessible. The principles discussed apply to data collected from other types of medical images (MRI, ultrasound, Positron emission tomography) with accommodations made for resolution of the scans produced by alternate modalities.

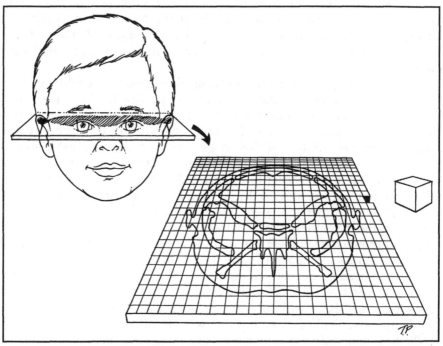

Figure 2.6 Pictorial representation of a single CT slice through the head of a child. The grid pictured is meant to represent the matrix of pixels (usually 256x256 or 512x512) that produces a picture of the anatomical features based on the linear attenuation coefficients that follow a scale having air at one extreme and a dense substances (e.g., mercury) at the other extreme. All tissues are assigned a score based on this scale. Pixels size, the dimension of a unit in the X and Y plane is used to produce a scale for the image. When the slices are combined to make a 3D reconstruction, the scale for the third dimensions, Z, is determined in part by the thickness of the slice images. A unit in the 3D matrix reconstructed in this way is called a voxel. An example of a voxel with X and Y units equal to pixel size and Z units equal to slice thickness is shown diagrammatically.

Landmarks from CT images

CT provides a two-dimensional representation of an anatomical section through the biological object. A useful metaphor is a loaf of bread that has been sliced. The pieces of bread are numbered consecutively starting at one end so that the relationship of one slice to the others and to the whole is known. You can remove a slice and see what the inside of the bread looks like anywhere along the loaf and replace the slice to reconstitute (reconstruct) the loaf.

When a biological form undergoes CT scanning, images are made that represent slices through the form. This is done by means of narrow collimated x-ray beams that penetrate the form and produce x-ray transmission measures. Large numbers of these measures are taken from many more beams targeted through the same cross section from many (approximately 200-1200) directions, some overlapping. The totality of these beam measurements are recorded and combined to reconstruct a two-dimensional matrix of coefficients that represent the linear x-ray attenuation coefficients of the scanned slice (Zonneveld, 1987). Each element of this matrix is referred to as a pixel, and each pixel has a known size that is used to produce a scale for the image. The reconstruction of all pixels produces a picture of the entire slice based on the linear attenuation coefficients that follow a scale having air at one extreme and a dense substance (e.g., mercury) at the other extreme (Figure 2.6). All tissues are assigned a score based on this scale.

Two-dimensional landmark coordinate data can be collected directly from slice images following methods similar to those discussed above for serially sectioned specimens. As long as the scale (pixel size) is known, 2D landmark locations can be recorded according to pixel column and row and later converted into metric units. In order to obtain a three-dimensional coordinate location for a landmark on a slice, information about the location of the point in the plane of the slice (X, Y coordinates), the thickness of the slice, and the location of that slice within the whole (Z coordinate) must be known. First, the slice containing a landmark of interest is identified. Next the X, Y coordinate of the landmark is recorded according to pixel column and row. The Z coordinate for each landmark is assigned according to table position (the position of the table for that particular slice as it moves through the x-ray tube), or it can be assigned according to a sequential numbering system from one end of the specimen to the other. The thickness of the slice must be accounted for in this numbering system. When viewing a single slice, all features exist on a single plane, so all land-marks within a slice will have the same Z coordinate. However, the software may determine whether these features are placed on the superior surface, the inferior surface, or projected midway between the upper and lower faces of the slice. When a feature appears in two con-tiguous slices, the Z coordinate of the corresponding landmark can be estimated as halfway between contiguous slices. If the thickness of the slice is greater than the dimensions of the pixel, the observer can expect measurement error to be disproportionately greater along the Z-axis as compared to the X- and Y-axes (see Richtsmeier et al., 1995).

Computerized reconstruction of CT images involves interpolation of the pixels between slice images, thereby producing a three-dimensional volume or surface of the form that has been scanned. There are several reconstruction programs available (e.g., etdips http://www.cc.nih.gov/cip/software/edtips/; Mullick and Venkatar-aman et al., 1998); ANALYZE (http://www.mayo.edu/bir/); (Robb and Hanson, 1995; Hanson and Robb et al., 1997; MEASURE (Barta and Dhingra et al., 1997), VOXBLAST (http://www.vaytek.com/VoxBlast.html). Some visualization and measurement software provide tools to determine the location of landmarks in 3D space. A landmark can be located simultaneously on the reconstruction, the axial images, and on recon-structions orthogonal to the axial plane (e.g., sagittal, coronal). These softwares and the hardware required to run them are becoming less expensive as personal computer technology advances.

2.4 Reliability of landmark coordinate data

Studying the reliability of the data collection methods is essential before any type of statistical analysis can be conducted. Moreover, a specific instrument may require certain types of testing. Our general advice is, know your instrument, talk with colleagues who have used the instrument previously, and design a measurement error study that will provide you with knowledge of the limitations of your data.

It is important to understand the various sources of error that can contribute to what is measured. Kohn and Cheverud (1992) define *pre-cision* as the average absolute difference between repeated measures of the same specimen. Precision of a method of data collection is mea-sured as the variability among repeated measures of the same specimen. Lack of precision results in variability among repeated mea-sures of the same specimen and has two components: 1) observer error in locating landmarks; and 2) instrument error in identifying landmark coordinates. Only specific research designs can partition measurement variability into these two components.

The study of repeatability, in contrast to precision, involves repeated measures taken on several specimens. Variability among repeated measures on more than one specimen includes variability due to the instrument, due to the observer, and due to biological differences among specimens. Repeatability is the precision of a particular measure relative to the biological differences among specimens, and it is measured as the ratio of the precision of a particular measure to the biological differences among specimens (Kohn and Cheverud, 1992).

This section deals only with the study of precision of various methods of data collection. The purpose is to isolate the effects due to the observer and due to the data collection instrument and not to confound it with the inherent biological variability among specimens. Biological variability is discussed in the next chapter.

Consider a single specimen fixed on a data collection device from which we are measuring landmarks. Suppose a single observer collects data from this specimen several times using a single instrument over a period of several days. In an ideal world, all coordinate measures would be identical to each other because the specimen is immobilized, but they usually are not. The variability in the measurements is due in part to instrument error and in part to error by the observer. Error studies should look for systematic differences in precision along the major axes.

Now suppose there is another observer who is also taking measurements on the same immobilized specimen using the same device. Though these measurements should be identical to the measurements made by the first observer, they rarely are. Differences in the two data sets stem from several sources. The second observer might be more or less experienced with the digitizer, or (s)he may know more or less about the specimen and the landmarks being collected. The differences between the measurements made by two or more observers of the same specimen constitute inter-observer error. Precision can be separately calculated for each investigator to determine the contribution of inter-observer error to the study. Excessive, inter-observer error can be avoided by limiting data collection to a single observer.

2.4.1 Measurement error studies

This section reports the measurement error associated with different data collection procedures by describing the results of studies conducted in our laboratory (Corner and Lele et al., 1992; Richtsmeier and Paik et al., 1995). By presenting two studies, we are emphasizing the fact that different data collection devices may require varying types of measurement error studies. The first study reports the measurement error associated with the 3Space digitizer (see Section 2.3.1). The second study discusses the measurement error associated with the landmark coordinates obtained from CT slice scan images (see Section 2.3.4).

> *a) Measurement error for collecting landmark coordinate data using the Polhemus 3Space digitizer*

Corner and colleagues (1992) studied the precision in landmark coordinate data collected by the Polhemus 3Space digitizer using the following data-gathering experiment (see Hildebolt and Vannier, 1988 for another measurement error study of this device). Two experienced observers collected coordinate data from 11 landmarks located on a single skull of a subadult male crab-eating macaque, *Macaca fascicularis*. The skull was secured to the digitizer table and digitized twenty times by each observer. Data were collected twice daily with approximately six hours separating the daily digitizations, until each observer's sample of 20 digitizations was complete. The time gap was introduced in order to ensure that the digitizations were independent of each other, reducing any "memory effect." Most importantly, the orientation and location of the specimen was fixed. Lack of movement of the object between digitizations is essential, as any movement of the object introduces random rotation and translation into measurement error calculations, making some of the variance parameters nonidentifiable (see Chapter 3 for the details on this issue). Having controlled for orientation and translation in this way, we can estimate the measurement error around each landmark. Measurement error is given by the sample variance for each landmark obtained from the 20 digitizations calculated using the coordinate locations for each axis. Since the data collectors were experienced, it is assumed that measurement error does not depend strongly on the data collector but depends mostly on the type of landmark and the measuring device.

    Table 2.1 (Corner and Lele et al., 1992) provides the measurement error associated with each of the landmarks along the three major axes. These are reported as standard deviations and not as variances (see Table 2.2 and Figure 1.2 for landmark identification). Each observer shows a high consistency in locating the landmarks. The standard deviation along any of the three axes ranges from 0.146 mm to 0.451 mm. The consistently low variation in coordinate values across the three dimensions indicates that no single direction is particularly prone to error for any of the landmarks considered here. Table 2.1 also indicates that Observer 2 (who was more experienced with the use of the machine and the anatomy of the skull) achieved a slightly better measurement error than Observer 1.

Table 2.1. Measurement error for 3Space digitizer (Corner et al., 1992). These are standard deviations (in *mm*) for the twenty repeated measurements of a fixed *Macaca fascicularis* skull. Larger values indicate larger measurement error or lack of precision. Definitions for the landmark abbreviations are given in Table 2.2 and some of the landmarks are pictured in Figure 1.2.

Landmark	Observer 1			Observer 2		
	X	Y	Z	X	Y	Z
NAS	.249	.379	.248	.173	.256	.195
NAL	.248	.451	.378	.361	.229	.304
ZMI	.163	.208	.338	.154	.143	.336
ZMS	.246	.385	.417	.199	.280	.446
IDS	.297	.395	.259	.287	.170	.205
PNS	.201	.281	.270	.228	.350	.373
MXT	.211	.180	.289	.300	.157	.446
VSJ	.226	.146	.253	.294	.166	.383
PTA	.224	.258	.343	.196	.216	.212
BRG	.165	.197	.204	.194	.219	.330
CAR	.200	.234	.239	.335	.201	.278

This study enables us to provide the following recommendations for the use of 3D digitizers as data collection devices. First, the researcher should become very familiar with both the specimens under study and the data collection device. We have found data collection practice sessions to be very useful. Second, when landmark data are collected but linear distances are computed for further statistical analysis (See

Table 2.2. Landmarks used in Corner et al. (1992) with their abbreviations

---

1) Nasion (NAS)
2) Nasale (NAL)
3) Right Zygomaxiilare inferior (ZMI)
4) Right Zygomaxiilare superior (ZMS)
5) Intradentale superior (IDS)
6) Posterior Nasal Spine:Vomer-Palatine intersection (PNS)
7) Right Maxilliary Tuberosity:Maxilliary-Palatine junction (MXT)
8) Vomer-Sphenoid junction (VSJ)
9) Right Pterion Anterior: Zygo-Spheno-Frontal junction (PTA)
10) Bregma (BRG)
11) Right Carotid Canal: center of canal (CAR)

Chapter 3 for such an approach), the impact of measurement error on the reliability of the distance data varies with the distance. The proportion of error is greater over small distances than it is over large distances. For example, in the measurement error study described above, the linear distances between any two landmarks ranged from 0.72 to 7.6 cms. Even if the maximum error along any of the axes (0.451 mm) occurred at a landmark defining an endpoint of the smallest linear distance, the error constitutes only 16% of that distance. For larger distances, the error is negligible. Consequently, the imprecision that can be tolerated in a study varies with the measures being used and comparisons being made (Kohn and Cheverud, 1992). Third, measurement error can systematically vary among the landmarks studied. For example, in this particular study larger measurement error was found at landmarks that mark the extreme ends of sutures located on an osseous edge (e.g., posterior nasal spine located at the distal end of the internasal suture; zygomaxillare superior located at the zygomaxillary suture along the orbital rim). Finally, if measurement error seems to be excessive, the following simple steps can be taken to reduce its impact. Data should always be collected at least twice. If neither trial contains large measurement error, the trials can be averaged to represent the specimen in analysis. When particular landmarks show increased error, and the landmark is needed in analysis, we suggest that the problematic landmark coordinates be collected repeatedly, preferably four to five times, from a fixed specimen and the average of the landmark coordinates be used for that specimen. Determining the number of repeated observations per specimen or per landmark depends primarily on two factors: 1) the distances between landmarks (smaller distances needing more repetitions); and 2) the landmark type (problematic landmarks requiring a larger number of repetitions).

b) Measurement error for collecting coordinate data from CT slice image

Use of computed tomography scans as a source for landmark coordinate data requires an alternate error study. A study conducted in our laboratory (Richtsmeier et al., 1995) evaluated the precision and repeatability of locating anatomic landmarks in three dimensions using CT slice images (each slice is 2D), and then validated these measures by comparing them to those taken using an already validated measurement system (the 3Space digitizer). We do not need to be concerned with issues of rotation and translation when comparing measures taken from a single set of images because once a specimen or patient undergoes a CT scan and the image is acquired, it is by definition fixed. However, we need to quantify any error that occurs in the acquisition of the image, and this requires that we compare data collected from various images of the same specimen.

In this study, 10 dry skulls underwent two episodes of CT scanning resulting in two sets of CT images for each skull. Landmarks were collected from each set of slice images during two separate data collection episodes. This resulted in four landmark sets being collected for each skull (two sets of landmark data from each image set). Differences in data collected from the two CT image sets of a single skull are due to error in digital recording by the CT scanner. Differences in data collected at separate times from the same image are due to recording error, which may have contributions from both the observer and the data collection software. This design enabled separation of error due to the imaging device from error in locating the landmarks on the images. Finally, the same landmark data were collected directly from the same ten skulls using the 3Space tabletop digitizer. Comparison of the landmark data sets collected from the CT scans and the digitizer quantifieserror in the way that data are acquired and displayed by a CT image and error in collecting data from the CT images (software).

The results of this study showed that three-dimensional landmark data collected from CT slice images are internally consistent and precise. It was noted, however, that error along the Z-axis (the axis along which the table is moving during image acquisition and which defines the slice thickness) is greater than error along the other two major axes. A subsequent study (Valeri et al., 1998) found error along the Z-axis to be reduced when the same data were collected from three-dimensional reconstructions of computed tomography slice images.

When collecting landmark data from 3D CT reconstructions, traditional landmarks that do not depend upon the orientation of the specimen are desirable because they exclude a potential source of error (Richtsmeier et al., 1995). Adding the coordinate system to the definition of a landmark adds another level of potential error to landmark identification. An example of a landmark that is dependent upon a specific orientation is glabella, defined as the most anterior projecting point on the frontal bone. The position of this landmark relative to other features of the skull will shift depending upon the orientation (coordinate system) of the specimen, and the orientation needs to be defined in order to collect comparable data from other specimens.

We use the term "local coordinate system" to refer to the coordinate system inherent to the object under study. When collecting landmark data from an object sitting on a digitizer, the local coordinate system is defined on the basis of the position of the object. If the object is moved, the local coordinate system changes. This is why an object must remain stable during data collection using a digitizer. However, when slice images or three-dimensional reconstructions of images (MRI or CT for example) are used for data collection, the local coordinate system is captured along with the image and remains inherent to the image regardless of any movement of the reconstructed image within most software programs. In effect, the exact coordinates for any given landmark (e.g., lambda) are the same whether collected from the 3D reconstruction oriented in the Frankfort horizontal position or from that same reconstruction turned upside-down and tilted at an oblique angle. The imaged specimen can be moved and tumbled through space during data collection.

We emphasize that these are only examples of measurement error studies for landmark data collected using specific devices. With each new instrument, researchers should conduct specialized studies of measurement error in order to assess the data collector's competency and the instrument error.

2.5 Summary

In this first part of Chapter 2, we have provided an outline of the types of data that could be used in morphometric analysis, a detailed explanation of the types of landmarks that could be identified for analysis, a summary of our experience with particular devices that can be used to collect landmark coordinate data, and guidance on potential sources of error. We stress that measurement error studies should always be done before analysis of data and that those measurement error studies must be designed with the data collection instrument in mind. In part 2 of this chapter, we provide a more thorough treatment of measurement error studies for landmark data. We provide the reader with those basics of matrix algebra that are required for managing landmark data, as well as an introduction to the concepts of rotation and translation and their effect on landmark coordinate data.

Statistical and Mathematical Preliminaries for Landmark Coordinate Data

We now introduce some statistical and mathematical concepts that are important for the statistical analysis of landmark data, whether conducting measurement error studies or comparing two forms or shapes. Matrix algebra is an essential mathematical tool in multivariate statistical analysis. It is particularly useful in the statistical analysis of landmark coordinate data. We strongly recommend that everyone read this part carefully. Readers who are mathematically and statistically challenged should at least try to familiarize themselves with the notation and some of the essential statistical and mathematical terminology introduced here, as it is used throughout the remaining chapters.

2.6 Introduction to matrix algebra

In this section, we introduce some of the basic ideas of matrix algebra. This is by no means a complete tutorial in matrices and their uses in multivariate analysis. For such a review, we refer the reader to Searle (1982). There are also many mathematical books on matrix algebra, e.g., Barnett (1990). The inquisitive reader may look to these sources for more details. The choice of topics covered here is dictated by our requirements for exposition of the use of matrix algebra in the analysis of landmark data.

   1) *Definition of a matrix:* A matrix is a rectangular array of elements arranged in rows and columns.

For example, $A = \begin{bmatrix} 1 & 2 & 3 \\ 4 & 5 & 6 \end{bmatrix}$ is a matrix.

2) *Dimension of a matrix:* A matrix with m rows and n columns is said to have dimensions $m \times n$ and is referred to as an $m \times n$ matrix. We read it as "m by n matrix". In text, we may either write an $m \times n$ matrix, or interchangeably, an m by n matrix.

The above matrix A is a 2 x 3 matrix. This matrix has 2 rows and 3 columns. Notice that this matrix has 6 elements. The number of elements in a matrix equals the product of the number of rows and the number of columns.

As in ordinary algebra, we use symbols to identify the elements of a matrix. In our applications the rows are represented by the symbol, i, while columns are represented by the symbol j.

$$
\begin{array}{ccc}
 & j=1 & j=2 \\
i=1 & a_{11} & a_{12} \\
i=2 & a_{21} & a_{22}
\end{array}
$$

An element is identified by its location in the matrix according to its row and column location. Thus, a_{ij} refers to the element in the i-th row and j-th column. For example, a_{12} is the element in the first row and second column. We read it as "a sub one two" or, if the context is clear, simply as "a one two". A commonly used short notation for a matrix is $A = [a_{ij}]_{i=1,2,...,m; j=1,2,...,n}$ or simply $A = [a_{ij}]$ if the range of the subscripts is obvious from the context. For notational convenience, we generally use the second notation.

For example, for the matrix, $A = \begin{bmatrix} 1 & 2 & 3 \\ 4 & 5 & 6 \end{bmatrix}$, the element $a_{13} = 3$

3) *Square matrix:* A matrix is said to be square if the number of rows equals the number of columns.

For example, $\begin{bmatrix} 1 & 2 & 3 \\ 4 & 5 & 6 \\ 7 & 8 & 9 \end{bmatrix}$ is a square matrix. This matrix

has 3 rows and 3 columns. Also notice that this has a total of 9 elements.

4) *Square symmetric matrix:* A square matrix of dimension $m \times m$ is said to be symmetric if and only if $a_{ij} = a_{ji}$ for all $i = 1,2,..,m;$ $j = 1,2,...,m$.

For example, $C = \begin{bmatrix} 0 & 1 & 1 \\ 1 & 0 & 1.41 \\ 1 & 1.41 & 0 \end{bmatrix}$ is a 3 x 3 square, symmetric matrix.

5) *Definition of a vector:* A matrix that has only one column and has dimension $m \times 1$, is called a vector. Sometimes this is also called a column vector. We denote a vector V of length m by $V = [v_i]_{i=1,2,...m}$.

For example, $V = \begin{bmatrix} 1 \\ 2 \\ 3 \end{bmatrix}$ is a vector of length 3. Also notice that, in this example, $v_1 = 1, v_2 = 2$ and $v_3 = 3$.

6) *Transpose of a matrix:* The transpose of a matrix is a matrix obtained by turning columns of the original matrix into rows while keeping their original order. The transpose of a matrix A is denoted by A^T.

For example, if $A = \begin{bmatrix} 1 & 2 \\ 3 & 4 \\ 5 & 6 \end{bmatrix}$ then $A^T = \begin{bmatrix} 1 & 3 & 5 \\ 2 & 4 & 6 \end{bmatrix}$

The transpose of a column vector is called a row vector.

7) *Equality of two matrices:* Two matrices A and B are said to be equal to each other if, and only if, they have the same dimension and $a_{ij} = b_{ij}$ for every i and j. That is, if and only if all of the corresponding elements are identical.

2.6.1 Addition and multiplication of matrices

Matrices do not behave like ordinary numbers. Addition and multiplication of matrices are allowed only if the matrices conform to certain dimensional restrictions.

1) *Addition of two matrices:* Let $A = [a_{ij}]$ and $B = [b_{ij}]$ be two matrices of the same dimension. Then $A + B = [a_{ij} + b_{ij}]$. The corresponding elements are added to obtain the sum of two matrices. The resulting matrix retains the same dimension.

Let $A = \begin{bmatrix} 1 & 2 \\ 3 & 4 \end{bmatrix}$ and $B = \begin{bmatrix} 5 & 6 \\ 7 & 8 \end{bmatrix}$.

Then $A + B = \begin{bmatrix} 1+5 & 2+6 \\ 3+7 & 4+8 \end{bmatrix} = \begin{bmatrix} 6 & 8 \\ 10 & 12 \end{bmatrix}$.

The addition of two matrices is allowed only if they have exactly the same dimensions.

2) *Multiplication of a matrix by a real number:* Let $A = [a_{ij}]$ be a matrix and c be a real number. Then $cA = [ca_{ij}]$. Multiplying a matrix by a real number results in the multiplication of each element by that real number. In mathematical literature, a single number (or a 1 x 1 matrix) is commonly referred to as a "scalar".

For example, $3B = 3 \times \begin{bmatrix} 5 & 6 \\ 7 & 8 \end{bmatrix} = \begin{bmatrix} 15 & 18 \\ 21 & 24 \end{bmatrix}$.

3) *Multiplication of a row vector by a column vector:* Let $V = [v_i]_{i=1,2,\ldots,n}$ be a row vector of dimension $1 \times n$ and $W = [w_i]_{i=1,2,\ldots,n}$ be a column vector of dimension $n \times 1$. Notice that V and W have the same number of elements. Let $U = VW$. Then U is given by $(v_1 w_1 + v_2 w_2 + \ldots + v_n w_n)$. Notice that this is a scalar (a single real number). A short notation for this sum that is often used is $\sum_{i=1}^{n} v_i w_i$.

As an example, let $V = \begin{bmatrix} 1 & 2 \end{bmatrix}$ and $W = \begin{bmatrix} 3 \\ 4 \end{bmatrix}$. Then $U = VW = (1 \times 3) + (2 \times 4) = 11$.

4) *Multiplication of two matrices:* Let A and B be two matrices. The product of these two matrices, $C = AB$, is defined only if the number of *columns* in matrix A is the same as the number of *rows* in matrix B.

Suppose A is a 4 x 3 matrix and B is a 3 x 5 matrix. Then AB can be calculated. However, given the condition stated above, BA cannot be calculated.

Let A and B be two matrices of dimensions $m \times n$ and $n \times p$. In this case, the product AB can be calculated. Let $C = AB$. Then the (i,j)-th element of the matrix C, denoted by C_{ij}, is given by the product of the i-th row of the matrix A and the j-th column of the matrix B. The matrix C is an $m \times p$ matrix.

Consider the matrices A and B, defined previously. The product of these two matrices is given by

$$AB = \begin{bmatrix} 1 & 2 \\ 3 & 4 \end{bmatrix} \begin{bmatrix} 5 & 6 & 9 \\ 7 & 8 & 10 \end{bmatrix} = \begin{bmatrix} 19 & 22 & 29 \\ 43 & 50 & 67 \end{bmatrix}$$

For example, the $(1, 1)$-th element of the product matrix is calculated by multiplying the first row of A, $\begin{bmatrix} 1 & 2 \end{bmatrix}$ by the first column of B, $\begin{bmatrix} 5 \\ 7 \end{bmatrix}$, using the multiplication of a row vector by a column vector described earlier. The other entries can be obtained similarly. The $(1,2)$-th element is obtained by multiplying the first row of A by the second column of B, the $(2,1)$-th entry is obtained by multiplying the second row of A by the first column of B, and the $(2,2)$-th element is obtained by multiplying the second row of A by the second column of B. Notice also that the resultant matrix is a 2 x 3 matrix. This dimension is given by the number of rows of the first matrix and the number of columns of the second matrix.

Exercise: Using the above rule and the vectors V and W defined above, show that $WV = \begin{bmatrix} 3 & 6 \\ 4 & 8 \end{bmatrix}$. Notice that WV is not the same as VW.

Notice also that this is a 2 x 2 matrix.

In general, for matrix multiplication, AB is not equal to BA, even when both products are well defined. Suppose A is an $m \times n$ matrix and B is an $n \times m$ matrix, then AB is an $m \times m$ matrix, whereas BA is an $n \times n$ matrix. By definition, if the two matrices are of different dimensions, they cannot be equal to each other.

We now turn to some simple applications of matrix algebra in the study of biological forms.

2.7 Matrix representation of landmark coordinate data

Landmark coordinate data, for both computational and statistical purposes, are most conveniently represented as a matrix. Suppose landmarks are collected from a three dimensional biological object. A typical "observation," a series of landmark coordinates obtained by the data collection methods described in Part 1, can be represented in a matrix form as described below.

For the sake of simplicity, instead of using proper biological names of the landmarks, we index them from 1 to K. It is assumed that a proper record of the correspondence between the biological nomenclature and the index is maintained consistently throughout the study.

The landmark coordinates of a three dimensional, K landmark object are written as a $K \times 3$ matrix as follows:

$$M = \begin{bmatrix} m_{11} & m_{12} & m_{13} \\ m_{21} & m_{22} & m_{23} \\ \vdots & \vdots & \vdots \\ \vdots & \vdots & \vdots \\ m_{K1} & m_{K2} & m_{K3} \end{bmatrix}$$

where the first row corresponds to the X, Y, and Z coordinates of landmark 1, the second row corresponds to the X, Y, and Z coordinates of landmark 2, and so on.

Thus M is a matrix with K rows and 3 columns with each row representing the X, Y, and Z coordinate values of the corresponding landmark. We say that M is a K by 3 matrix. M is called a *landmark coordinate matrix*. For a two-dimensional object where only X and Y coordinates exist for each landmark, the landmark coordinate matrix is a K by 2 matrix.

Throughout this monograph, we use the letter K to denote the number of landmarks, and letter D to denote the number of dimensions of the object, unless specified otherwise. The dimension D is typically either 2 or 3. In general, a landmark coordinate matrix is a K by D matrix of real numbers.

The next question we ask is: what happens to the landmark coordinate matrix of a given object if we rotate (spin around a point) or translate (shift) the object? The rotation and translation of objects can be described using matrix algebra. This requires a few more definitions.

a) *An identity matrix*: An identity matrix is a square matrix with diagonal elements equal to 1 and all the other elements equal to 0. An identity matrix is denoted by I.

For example, $\begin{bmatrix} 1 & 0 & 0 \\ 0 & 1 & 0 \\ 0 & 0 & 1 \end{bmatrix}$ is a 3 by 3 identity matrix. In this monograph we will denote such a matrix by I_3 which tells us that the matrix is an identity matrix of a specific dimension, in this case 3. Thus I_D is a D by D identity matrix. An important property of an identity matrix is that if we multiply any matrix by an identity matrix, the resultant matrix is the same as the original matrix.

b) *An orthogonal matrix*: A square matrix is an orthogonal matrix if the product of itself and its transpose is an identity matrix. In other words, a matrix R is an orthogonal matrix, if $RR^T = R^T R = I$.

Exercise: Verify that a matrix $\begin{bmatrix} \cos(\theta) & -\sin(\theta) \\ \sin(\theta) & \cos(\theta) \end{bmatrix}$ is an orthogonal matrix. To verify this result, recall that $\sin^2(\theta) + \cos^2(\theta) = 1$.

Because this matrix features in the rotation of an object by an angle θ, we denote it by $R(\theta)$. Thus, we write

$$R\theta = \begin{bmatrix} \cos(\theta) & -\sin(\theta) \\ \sin(\theta) & \cos(\theta) \end{bmatrix}.$$

We refer to $R(\theta)$ also as the rotation matrix or the rotation parameter.

c) *Rotation of an object*: Let the landmark coordinate matrix for a given two-dimensional object be denoted by M. Suppose we rotate this object by an angle θ. The landmark coordinate matrix corresponding to the rotated object, $\tilde{M}$, can be obtained by multiplying the original landmark coordinate matrix by the orthogonal matrix $R(\theta)$. Thus, we get $\tilde{M} = MR(\theta)$.

Similarly, for a three dimensional object one can obtain the landmark coordinates of the rotated object by multiplying the original landmark coordinate matrix by a 3 by 3 orthogonal matrix. A 3 by 3 orthogonal matrix is given by:

$$\begin{bmatrix} \cos(\alpha)\cos(\phi) & \sin(\alpha)\cos(\phi) & \sin(\phi) \\ -\sin(\alpha) & \cos(\alpha) & 0 \\ -\cos(\alpha)\sin(\phi) & -\sin(\alpha)\sin(\phi) & \cos(\phi) \end{bmatrix}.$$

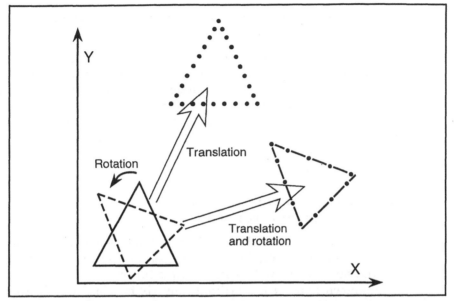

Figure 2.7 Pictorial representation of the operations of translation and rotation. The triangle drawn with an unbroken line is the original configuration. The triangle drawn with a broken line represent a change in the triangle due to rotation on an axis with no translation. The triangle drawn with a dotted line represents a change in the triangle due to translation across the plane with no rotation. The triangle drawn with dots and dashes represents the result of the combination of rotation and translation.

Notice that rotation of a three-dimensional object requires specification of two angles of rotation α and ϕ.

d)   *Translation of an object*: Let the landmark coordinate matrix for a given two dimensional object be denoted by M. Suppose we translate this object so that all the X-coordinates are moved by t_1 units and the Y-coordinates are moved by t_2 units. The landmark coordinate matrix of the translated object is given by $M + \underline{1}t$ where $\underline{1} = \begin{bmatrix} 1 \\ 1 \\ \vdots \\ 1 \end{bmatrix}$ a $K \times 1$ column vector of 1's and

t is a 1 by 2 row vector. We call "t" the translation vector or a translation parameter.

For a three-dimensional object, the translation vector $t = [t_1\ t_2\ t_3]$ is a 1 by 3 vector.

One can combine the two operations and obtain the landmark coordinate matrix of a translated and rotated object by $MR(\theta) + \underline{1}t$. Figure 2.7 provides the pictorial representation of these three operations. Figure 2.7 provides the original landmark coordinates for the triangle, a rotation (R) of the original, a translation (t) of the original, and the combination of rotation and translation ($R + t$) of the original.

2.8 Statistical model and inference for the measurement error study

In the measurement error study described in the previous part of this chapter, we fixed a single skull to the digitizer and collected the coordinates of the same set of landmarks multiple times. Let M_i denote the true but unknown landmark coordinate matrix for this skull in this fixed orientation. Each data collection episode produced a landmark coordinate matrix, say M_i, where "i" denotes the i-th data collection episode. Due to measurement error, we will not get exactly the same landmark coordinate values during every data collection episode. The variability among the M_i's denotes the measurement error. We need a statistical model to study and quantify this variability.

2.8.1 Preliminaries of the matrix normal distribution

First notice that the M_i's are related to the true landmark coordinate matrix M. This relationship may be written as $M_i = M + E_i$. That is, M_i is obtained by adding an error matrix E_i to the true coordinate matrix M. We assume that this error matrix has certain properties. These assumptions are detailed below.

Assumption 1: There is no systematic bias introduced when locating a landmark in space. That is, the measurement errors at any particular landmark along any particular direction cancel each other out and are on an average zero.

Assumption 2: The errors introduced at any particular landmark and along any particular direction are distributed according to a Normal distribution. Though normally distributed, these errors, however, may be correlated with each other.

Errors local to a landmark and the association of errors between landmarks and along directions are calculated as variances and covariances and can be collected in a matrix form. For the definitions of variance and covariance, the reader may refer to any elementary statistics textbook, (e.g., Rosner, 1995).

Let us start with a simple situation of three landmarks on a two dimensional object. In this case, the true landmark coordinate matrix M is a 3 by 2 matrix. There is measurement error associated with each landmark along each direction.

Focusing only on the X direction, let us define

$$V_X = \begin{bmatrix} V_{X,11} & V_{X,12} & V_{X,13} \\ V_{X,12} & V_{X,22} & V_{X,23} \\ V_{X,13} & V_{X,23} & V_{X,33} \end{bmatrix}$$

where $V_{X,11}$, $V_{X,22}$, and $V_{X,33}$ are the variances representing measurement errors in the X direction for landmarks 1, 2, and 3 respectively; $V_{X,12}$ is the covariance between errors in the X direction for landmarks 1 and 2, $V_{X,13}$ is the covariance between errors in the X direction for landmarks 1 and 3, and $V_{X,23}$ is the covariance between errors in the X direction for landmarks 2 and 3. Notice that the matrix V_X is a square, symmetric matrix, and that there are only six distinct entries. Similarly, we can write the variance matrix V_Y for the errors in the Y direction. The covariance matrix representing errors between the X and Y directions is written as $V_{X,Y}$. Each of these is a 3 by 3 square, symmetric matrix. These matrices can be combined to form a new 6 by 6 matrix,

$$V = \begin{bmatrix} V_{X,11} & V_{X,12} & V_{X,13} & V_{XY,11} & V_{XY,12} & V_{XY,13} \\ V_{X,12} & V_{X,22} & V_{X,23} & V_{XY,12} & V_{XY,22} & V_{X,23} \\ V_{X,13} & V_{X,23} & V_{X,33} & V_{XY,13} & V_{XY,23} & V_{XY,33} \\ V_{XY,11} & V_{XY,12} & V_{XY,13} & V_{Y,11} & V_{Y,12} & V_{Y,13} \\ V_{XY,12} & V_{XY,22} & V_{XY,23} & V_{Y,12} & V_{Y,22} & V_{Y,23} \\ V_{XY,13} & V_{XY,23} & V_{XY,33} & V_{Y,13} & V_{Y,23} & V_{Y,33} \end{bmatrix}.$$

This matrix is sometimes written in a shorter form as $V = \begin{bmatrix} V_X & V_{X,Y} \\ V_{X,Y} & V_Y \end{bmatrix}$.

Notice that the first three diagonal elements of this matrix correspond to the variance in the X direction for landmark 1, 2, and 3 respectively. Similarly the next three diagonal elements of this matrix correspond to the variance in the Y direction for landmark 1, 2, and 3 respectively. Other elements (called the off-diagonal elements) provide information on the covariances. Such a matrix V is called either a "variance-covariance matrix," a "covariance matrix," or a "variance matrix." All three names are acceptable. In our application the matrix V summarizes the measurement errors. This matrix V is a $2K \times 2K$ matrix where K, the number of landmarks is 3 and the dimension of the object is 2.

Remark: Let us refer to Table 2.1. In that particular measurement error study, we have 11 landmarks and the *Macaca fascicularis* skull is a three-dimensional object. The resultant covariance matrix is a 33 by 33 symmetric matrix. This table provides only the square root (standard deviations) of the diagonal elements of the covariance matrix. For example, the first entry in the table, 0.249, is equal to $\sqrt{V_{X,11}}$, the standard deviation of the measurement error for landmark 1 along the X direction. The remaining entries in the column include the values of $\sqrt{V_{X,22}}$, $\sqrt{V_{X,33}}$, and so on.

Exercise 1) Verify that if the object under study is two-dimensional and has K landmarks, the variance-covariance matrix V is a $2K \times 2K$ square, symmetric matrix.

Exercise 2) Verify that if the object under study is three-dimensional and has K landmarks the variance-covariance matrix V is a $3K \times 3K$ square, symmetric matrix. Notice that for three dimensional objects,

$$V = \begin{bmatrix} V_X & V_{X,Y} & V_{X,Z} \\ V_{X,Y} & V_Y & V_{Y,Z} \\ V_{X,Z} & V_{Y,Z} & V_Z \end{bmatrix} .$$

The observations just presented, that the true landmark coordinate matrix is M, that measurement errors E_i are Normally distributed with variance covariance matrix V, and that $M_i = M + E_i$, are summarized by the notation: $M_i \sim N(M,V)$. We read this as "M_i follows a matrix Normal distribution with mean matrix M and the variance-covariance matrix (or simply variance) V."

Remember that for a two dimensional object M is a K by 2 matrix, and for a three-dimensional object, it is a K by 3 matrix. The variance matrix V is a square, symmetric matrix of dimension $2K \times 2K$ when the dimension of the object is two, and $3K \times 3K$ when the dimension of the object is three. When the object is three dimensional, the first K diagonal elements of V quantify variability in the X direction, the next K elements quantify the variability in the Y direction, and the last K elements quantify the variability in the Z direction for landmarks 1 through K. In the measurement error study we concentrate on the diagonal elements (variances) and ignore the off-diagonal elements (covariances). The matrix V can be used to identify those landmarks that are highly error prone.

In practice, the matrix V is unknown, and we have to use the observations from N data collection episodes to estimate it. Let $M_1, M_2,...M_N$ denote the observations (the landmark coordinate matrices) from N data collection episodes.

We start with calculating the mean of these observations. Let $\overline{M} = \frac{1}{N}\sum_{i=1}^{N} M_i$ denote the sample mean matrix. Notice that $\overline{M}$ is a

K by 2 or a K by 3 matrix, depending on the dimension of the object.

Let A be an $m \times n$ matrix. Then $A^{<i>}$ denotes the i-th column of A. For example, $M_i^{<2>}$ denotes the second column of the observation M_i. With this notation, for a two-dimensional object, the variance-covariance matrix V is obtained using the following formula. The first equation provides the variance-covariance matix in the X direction, the second equation provides the variance-covariance matrix in the Y direction, and the third one provides the covariances between X and Y directions.

$$V_X = \frac{1}{N}\sum_{i=1}^{N}(M_i^{<1>} - \overline{M}^{<1>})(M_i^{<1>} - \overline{M}^{<1>})^T$$

$$V_Y = \frac{1}{N}\sum_{i=1}^{N}(M_i^{<2>} - \overline{M}^{<2>})(M_i^{<2>} - \overline{M}^{<2>})^T$$

$$V_{X,Y} = \frac{1}{N}\sum_{i=1}^{N}(M_i^{<1>} - \overline{M}^{<1>})(M_i^{<2>} - \overline{M}^{<2>})^T$$

Extension to three-dimensional objects is straightforward. The additional terms needed are given by:

$$V_Z = \frac{1}{N} \sum_{i=1}^{N} (M_i^{<3>} - \overline{M}^{<3>})(M_i^{<3>} - \overline{M}^{<3>})^T$$

$$V_{X,Z} = \frac{1}{N} \sum_{i=1}^{N} (M_i^{<1>} - \overline{M}^{<1>})(M_i^{<3>} - \overline{M}^{<3>})^T$$

$$V_{Y,Z} = \frac{1}{N} \sum_{i=1}^{N} (M_i^{<2>} - \overline{M}^{<2>})(M_i^{<3>} - \overline{M}^{<3>})^T$$

Strictly speaking, the above formulae provide an *estimate* of the variance covariance matrix. For the sake of simplicity of exposition, in this chapter, we de-emphasize the difference between the true V and its estimated value.

2.8.2 Further properties of the matrix valued normal distribution

Chapter 3 introduces statistical models to model variability between different individuals within a population. We call such variability "the biological variability." The following concepts and notation are useful to understand the discussion in Chapter 3. In particular, we introduce a special form for the variance-covariance matrix based on the Kronecker product which is mathematically convenient and biologically reasonable for modelling biological variability. We also introduce the idea of transformation of a random variable and discuss the impact of such transformation on the distribution of the random variable.

Kronecker product of two matrices
Let A be an $m \times n$ matrix and B be a $p \times q$ matrix. Then the Kronecker product of A and B is defined as:

$$\text{Then } A \otimes B = \begin{bmatrix} a_{11}B & a_{12}B & \cdots & a_{1n}B \\ a_{21}B & a_{22}B & \cdots & a_{2n}B \\ \vdots & \vdots & \vdots & \vdots \\ a_{m1}B & a_{m2}B & \cdots & a_{mn}B \end{bmatrix}$$

The resultant matrix is an $mp \times nq$ matrix.

For example, let $A = \begin{bmatrix} 1 & 2 \\ 3 & 4 \end{bmatrix}$ and $B = \begin{bmatrix} 1 & 2 & 3 \\ 4 & 5 & 6 \\ 7 & 8 & 9 \end{bmatrix}$.

$$\text{Then } A \otimes B = \begin{bmatrix} 1 & 2 & 3 & 2 & 4 & 6 \\ 4 & 5 & 6 & 8 & 10 & 12 \\ 7 & 8 & 9 & 14 & 16 & 18 \\ 3 & 6 & 9 & 4 & 8 & 12 \\ 12 & 15 & 18 & 16 & 20 & 24 \\ 21 & 24 & 27 & 28 & 32 & 36 \end{bmatrix}.$$

Exercise: Calculate $B \otimes A$. Show that it is not equal to $A \otimes B$.

Also, notice that the Kronecker product can be calculated for any two matrices. Recall that in order to calculate the usual product of two matrices, the number of columns of the first matrix has to be equal to the number of rows of the second matrix.

Modeling the variance-covariance matrix V for the matrix normal distribution

Let us consider a two-dimensional object with K landmarks. We know that the variance-covariance matrix corresponding to the measurement error model described above is of dimension $2K \times 2K$. In some special situations, the matrix V can be written in the Kronecker product form. We describe some of these situations below. In the next chapter, we argue that these models are mathematically convenient and biologically reasonable for modeling biological variability.

> *Situation 1:* Suppose the measurement errors at all landmarks and along all directions are identical to each other. In addition, suppose that they are also uncorrelated to each other. In this situation the variance-covariance matrix V has the same quantity, say σ^2, along the diagonal and all off-diagonal entries are zero. Such a matrix may be written as $\sigma^2 I_{2K}$. It can also be written equivalently in the Kronecker product form as $\sigma^2 I_K \otimes I_2$. This model is sometimes known as the "isotropic errors" model.

> *Situation 2:* The assumption of isotropic errors may seem unrealistic in practice. Suppose the measurement errors at different landmarks are correlated, but that the error along different axes are uncorrelated and equal. The covariance

matrix corresponding to such a measurement error structure can be written as $\Sigma_K \otimes I_2$ where Σ_K is the covariance matrix between landmarks. Remember that in order for Σ_K to be a covariance matrix, it must be a $K \times K$ square symmetric matrix.

Situation 3: In the above model, we may want to include different magnitudes of variability along the two axes and also correlations between the axes. In this case, the covariance matrix may be written as $\Sigma_K \otimes \Sigma_2$ where Σ_K is a $K \times K$ square, symmetric matrix, and Σ_2 is a 2 by 2 square, symmetric matrix.

Remarks:

1. These examples of covariance matrices for measurement error are not exhaustive. There are certain covariance structures that are not captured by the above matrices.
2. Generalization to three-dimensional object entails replacing all 2 x 2 matrices in the above discussion by 3 x 3 matrices.
3. Strictly speaking, the matrices Σ_K and Σ_D should also be positive semidefinite, that is, all the eigenvalues of these matrices should be non-negative. See Barnett (1990) for details on positive semidefinite matrices.

2.8.3 Effect of rotation and translation on the matrix normal distribution

The following observations are standard results from the theory of matrix valued Normal distribution. See Arnold (1981, page 312) for mathematical details. These results are used extensively in Chapters 3 and 4. For this reason, we present them here. Readers who are not comfortable with the mathematical aspects should at least read through this part and familiarize themselves with the notation and the intuitive meaning of these results. This will prepare the reader for the less-mathematical discussion found in Chapters 3 and 4, Part 1.

Let us first try to understand *intuitively* what we mean by the statement an observation X, "comes from" or equivalently, "follows" a normal distribution. Let us say that X corresponds to "height of a person." Consider the population of all individuals in which we are interested. Suppose we calculate the mean of the heights of all these individuals. This mean is called the "population mean," usually denoted by μ (*mu*). Similarly the variance of the heights of all these individuals is called the "population variance," denoted by σ^2 (*sigma square*). Suppose further that, when we plot the histogram of the heights of all the individuals in the population, it looks like a "bell shaped curve." Then we say that the population distribution is Normal. Now suppose we select an individual from this population at random. What can we say about the height of this individual? The height of this individual is more likely to come from the "middle" of the distribution near μ than the tails. In statistical terminology, this is expressed as $X \sim N(\mu, \sigma^2)$ and we read it as "X follows a normal distribution with mean μ and variance σ^2." Suppose we select "n" different individuals randomly from this population. Let X_i denote the height of the i-th individual. We say that $X_i \sim N(\mu, \sigma^2)$ are independent (because the individuals are selected randomly from the population) and identically distributed (because they come from the same population) random variables.

Let us now say that the experimenter used inches as the unit for measuring height. The population values, also called the parameters, μ and σ^2, are in the units of inches and inches^2 respectively. Suppose that by mistake the height of the 10th individual was measured and reported in units of centimeters. Clearly we cannot say that $X_{10} \sim N(\mu, \sigma^2)$ because the units are different. What we can say is that: $X_{10} \sim N(2.5\mu, (2.5\sigma)^2)$. We simply change the units for the mean parameter and the variance parameter to match the units of the other measurement. The moral of the story is that if we transform an observation, the distribution of that observation changes as well. We next generalize the idea of transformation to Matrix Normal distribution and landmark coordinate matrices.

Recall that a matrix Normal distribution is characterized by two quantities, the mean matrix M and the variance-covariance matrix V. Suppose that the covariance matrix V can be written in the Kronecker product form, $V = \Sigma_K \otimes \Sigma_D$. Suppose we observe a landmark coordinate matrix of an object and denote it by X. Suppose further that $X \sim N(M, \Sigma_K \otimes \Sigma_D)$. Below we summarize how translation and rotation of the original object affects the distribution. (This is similar to the transformation from inches to centimeters described in the previous paragraph.)

a) Effect of translation: $X + \underline{1}t \sim N(M + \underline{1}t, \Sigma_K \otimes \Sigma_D)$. This means that if we translate an object, the mean matrix gets translated by the same amount, but the variance-covariance matrix remains unaffected.

b) Effect of rotation: $XR \sim N(MR, \Sigma_K \otimes R^T\Sigma_D R)$. This means that rotation of an object affects the mean as well as the variance-covariance matrix.

c) Effect of translation and rotation combined: $XR + \underline{1}t \sim N(MR + \underline{1}t, \Sigma_K \otimes R^T\Sigma_D R)$. This means that rotation and translation of an object affects both the mean and the variance-covariance matrix.

CHAPTER 3

Statistical Models for Landmark Coordinate Data

In this chapter, we present models that are used to study form and variability among forms as represented by landmark data. In Part 1 of this chapter, we discuss the concept of variability among organisms, models that are used to characterize variability and the way in which these models relate to the methods used for analysis of form. Based on these considerations, we then discuss the landmark coordinate representation of form and present our method for characterizing form without reference to a coordinate system. The final portion of Part 1 of this chapter presents analysis of the example data sets presented in Chapter 1 using these methods and models. Part 2 of this chapter presents the formal statistical theory for characterizing a single population of forms using landmark data and presents the necessary computational algorithms for estimation of the mean and variance.

Before we discuss models and methods for the analysis of landmark data, it is necessary to differentiate between a statistical model and a statistical method. A model, as used in this monograph, is a mathematical construct that attempts to characterize certain aspects of the underlying phenomenon. This mathematical construct includes quantities called parameters. For example, consider the simple linear regression model that describes the relationship between height and weight of an individual. Let Y denote the weight of an individual and X denote the height of the same individual. The statistical model used in this situation may be: $Y = \beta_0 + \beta_1 X + \varepsilon$. This model states that the weight of an individual is a linear function of the height of the individual with the addition of some random variability. The parameters of this model are the intercept, β_0, and the slope, β_1. The term ε denotes the variability around the mean response $\beta_0 + \beta_1 X$. Usually, we assume that the variability around the mean can be modeled by a Normal distribution with mean 0 and variance σ^2. We also assume that the individuals are independent of each other. For this particular case, this assumption means that the individuals are not siblings or otherwise related.

The simple statistical model presented for the relationship between height and weight includes all of the following considerations: linearity of the relationship, Normal variability, and individuals that are independent of each other. The validity of each component of the model can be debated. If a component is changed, a new statistical model is obtained.

Notice that data do not enter into the discussion of the model. A model is formulated using the intuition of the scientist and whatever previous experience and knowledge (s)he may have. Once a model is formulated, the data are used to determine those parameters of the model that are most compatible with the observations. This process is called "estimation of the parameters." Many different methods may exist for the estimation of the parameters. A method is any technique used in the analysis of data.

While conducting any scientific study, a model should be specified *first*. Following the specification of a model, methods should be devised or chosen for estimating the parameters of the model and for conducting any other relevant data analysis. A particular method is judged as good or bad, correct or incorrect, only in relation to its efficacy under a particular model. An evaluation of the conclusions of any scientific analysis of data must consider the validity of the model *and* the correctness of the method.

3.1 Statistical models

"The ubiquity of noise is why statisticians and statistics are useful to science — if there is no noise, no uncertainty, then there is no need for us or it."

Holland (1992)

There is variation among biological organisms. No two individuals, no matter how closely related, are exactly the same. Even if two individuals are twins or clones, variability between the phenotypes exists due to the individual's interactions with the environment. To describe this variability and to enable proper statistical comparison of groups, a method for estimating variability within a population is needed.

The first and most important step in the statistical analysis of landmark data is the specification of a model for variability within a group. This step is important because the model that is adopted and used in analysis influences the analytical results. An inappropriate model can produce illusory or incorrect results. As discussed above, we use the term model to describe the mathematical construct that attempts to characterize (not calculate) how individuals within a population vary. Unfortunately, there is no such thing as the correct model. All models, even highly complex ones, are approximations.

The choice of a suitable model is dependent on several factors including previous experience of the investigator, knowledge of the specimens, familiarity with the nature of the data, and mathematical convenience. In choosing a model, we need to be concerned that it is realistic enough to model observable facts reasonably well, but at the same time simple enough mathematically to allow statistical inference. The parameters of the chosen statistical model (e.g., the mean and the variance) should help elucidate the biological phenomena, and should be estimable using a sample of observations. For example, if technical problems do not allow the collection of three-dimensional data from biological organisms that are truly three-dimensional, the observer needs to make an informed decision regarding whether the organisms can be adequately represented by two-dimensional data. Additionally, any model of the variability of landmark coordinate data should be such that all the observations generated under such a model belong to the same dimensional space as the original data (i.e., either two-dimensional or three-dimensional Euclidean space).

In short, three criteria should be met in designing the model and the parameters: 1) the model should be a reasonable approximation of the variability within a population; 2) the parameters should convey a direct, tangible meaning to the phenomena under study (i.e., a biological meaning for biological data sets); and 3) the parameters should be estimable using available data. The first two criteria require knowledge of the biology of the specimens. The third criterion requires statistical knowledge. The aim of this chapter is to suggest a reasonable statistical model for landmark coordinate data and to discuss the estimation of the relevant parameters.

3.2 Model for intra-group variability

Our model of intra-group variability can be illustrated by conducting the following experiment. Imagine that you have eleven transparencies in front of you. Assume that these transparencies are not the familiar rectangular ones, but instead are circular, each with a different radius. The importance of the circular transparencies will become evident as you work through the experiment. On one transparency, draw a triangle (using a red pen) and label each of the three vertices as a landmark using the numbers 1, 2, and 3. Take a second transparency and put it on top of the first. Using a black pen, place a single point in the proximity of the original (red) landmark 1 at a random distance and direction from landmark 1. Next, on the same transparency, place a point in the proximity of landmark 2 at a random (though small) distance and direction from the original landmark 2. Do the same for landmark 3. You now have a three-point configuration on the second transparency that was created using the original red triangle as a template. Repeat this procedure of forming a novel three point configuration in black ink nine more times, using a different transparency each time. At completion, you will have a stack of eleven transparencies. Ten of these contain a random three-point configuration drawn in black that are related to the original three point configuration (drawn in red). The spread of the black points around each of the red points can be thought of as the variability around an average location.

Biological organisms that constitute a group resemble each other to such a degree that we have an intuitive understanding of a typical or "average" form (the red triangle) that represents all members of the group. Individuals within a group correspond to the ten black triangles that were just created on the circular transparencies. Some members are very similar to the average, while others are less like the average. Since all forms differ from each other in various ways, a scheme for characterizing these differences is needed. It is convenient to organize and specify these differences as divergences from an average form.

In biology, genetic and environmental influences combine to affect structures represented by each of the landmarks, thereby creating randomly perturbed forms. When observing a group of forms, we think of these individuals and the variation among them by relating them to a typical form that does not exist but that we are able to envision. This is formally done in statistics by calculating an average form and measuring variation in the sample with reference to an average.

To quantify variability within a sample, the researcher needs to make some decisions about how the individuals in a sample vary. These decisions constitute the choice of a *perturbation model*, the mathematical construct that characterizes the way in which individuals within a population vary with respect to an average form. The degree to which a specific model correctly approximates the perturbed copies of the average form depends upon the appropriateness, or the "fit" of the model to the data. The better we are able to anticipate these perturbations, the better the model will fit the data. Again, knowledge of the study specimens puts the researcher in a better position to choose the appropriate model.

Suppose we assume that a normal distribution will model the perturbation of the transparency data appropriately. Let M denote the mean form. M can be likened to the notion we have of a form that has the typical features of individuals within the group and that fairly represents members of a particular group. In the transparency experiment M corresponds to the red triangle. Since we are dealing with landmark data, M is a $K \times D$ matrix where K corresponds to the number of landmarks and D corresponds to the dimension of the form. Each of the perturbed observations (black triangles) are obtained by adding some noise, namely E_i, to the mean form, where E_i's are assumed to be matrix-valued normal random variables with a mean of 0 and a covariance matrix. Each observation is written as $M + E_i$. In our transparency experiment, each $M + E_i$ corresponds to one of the ten configurations drawn using the black pen.

In a simple world, these would be our only concerns in choosing a model. However, there is an additional complication. This complication involves the relationship between the coordinate system local to the object and the coordinate system used to collect the landmark coordinate data. No information is available regarding how these coordinate systems relate to one another. To clarify this problem, imagine that after creating the ten transparencies, they are dropped onto the floor. When we pick the transparencies up and put them onto the table, their original orientations are changed. Because they are circular, we have no outside reference to inform us about how they were related to one another before they fell to the floor. In order to study biological variability, knowledge of the relationship between the transparencies before they fell to the floor is essential.

Let us now go back to the original, undisturbed stack of transparencies and consider the effect of lifting a single transparency from its original position and putting it elsewhere on the table. Movement of a single transparency introduces a particular translation and rotation to the definition of this transparency in relation to its original location. Lifting the next transparency and moving it to another location on the table introduces a different translation and rotation specific to that transparency.

Rotation refers to change in orientation characterized as movement around an axis (Figure 2.7). Imagine placing a pin through the original pile of transparencies. This pin can be used as an axis about which the black transparencies spin. Upon rotation, the relative locations of the points on any single transparency remain the same, but the orientation of the perturbed observations (the triangles) changes by rotation around an axis. Mathematically, rotation of an object corresponds to multiplication of the landmark coordinate matrix by an orthogonal matrix (see Chapter 2, Part 2).

Translation, on the other hand, corresponds to a black transparency remaining stable in terms of rotations around axes but sliding in any direction along the plane defined by the tabletop (Figure 2.7). As with rotation, the relative locations of points are maintained under translation. Mathematically, translation corresponds to adding a matrix of identical rows to a matrix (see Chapter 2, Part 2).

With these definitions of rotation and translation in hand, we can now present the data using the full perturbation model which represents the observations after they have been arbitrarily translated and rotated. A single observation, X_i, incorporating the full perturbation model is represented mathematically by: $X_i = (M + E_i)R_i + \underline{1}t_i$ where E_i is the random perturbation of the mean form M, R_i is the orthogonal matrix corresponding to rotation of X_i, and t_i is the translation matrix. We emphasize that each observation, X_i, may be rotated and translated differently. In the context of the transparency experiment, X_i corresponds to the landmark coordinate matrix representing the i-th transparency after it has been dropped onto the floor, picked back up, and put onto the table. The original landmark coordinate matrix for any black transparency (before disturbing it) is given by $M + E_i$. Once the matrix has been arbitrarily rotated and translated, it is written as $(M + E_i)R_i + \underline{1}t_i$.

Let us now deal with modeling the perturbation pattern that was used to generate the black transparencies from the red one. Perturbation patterns are quantified by the covariance structure of the random variables, E_i. These covariance matrices are best conveyed by a graphic representation of the variability implied by different covariance structures. Let us begin by accepting, for instructional purposes and mathematical convenience, the assumption of Gaussian (Normal) perturbations across all landmarks on a two-dimensional, three-landmark object. Complex objects with more landmarks in three dimensions can be modeled using similar covariance matrices.

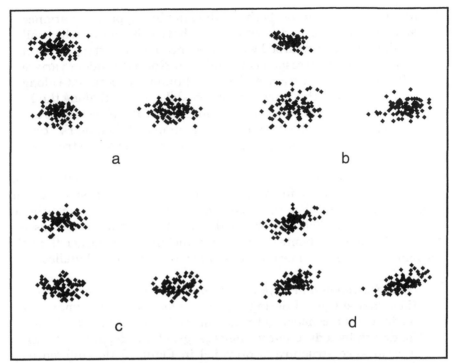

Figure 3.1. Four types of covariance structures that may be used in the matrix normal perturbtion model: a) independent and same variances (isotropic); b) independent and different variances; c) correlated landmarks and uncorrelated axes; d) correlated landmarks and correlated axes. We generated 100 matrix Normal random variables under different covariance structures and plotted these in the original coordinate system without introducing any rotation, translation or reflection. Visually, it is difficult to differentiate the first three covariance structures. However, the statistical inferences estimated from these different covariance structures (e.g., confidence intervals) are quire different under these different covariance structures. The mathematical details of the various covariance structures conveyed graphically in Figure 3.1, are provided in Part 2 of this chapter.

We will consider four different types of covariance structures (graphically presented in Figure 3.1), chosen for both statistical and biological reasons. Statistically, these four types of covariance structures are amenable to analysis using a sample of arbitrarily rotated and translated objects. Biologically, these covariance structures seem reasonable and appropriate, reflecting expectations gained through experience with data collected from biological populations. Mathematical details of the various covariance structures conveyed graphically in Figure 3.1, are provided in Part 2 of this Chapter.

(a) Isotropic variance. Perhaps the simplest type of covariance structure for the perturbation distribution is one in which all landmarks are perturbed with the same variance irrespective of direction. This corresponds to adding a Normal random variate with mean 0 and variance σ^2 to each landmark coordinate along both the X- and the Y-axes. In this model σ^2 is equal along the X- and Y-axes. Figure 3.1a shows a collection of 100 objects that are generated from the template of a simple 3 landmark, two-dimensional object using this perturbation covariance structure.

This isotropic variance structure is easy to visualize, but may not be biologically realistic in the study of certain biological structures or certain populations. For example, we found in a sample of primate skulls (Lele and Richtsmeier, 1990) that the variances for osseous landmarks that mark sites of muscle attachment are larger than the variance of those that mark foramina for neurovascular bundles.

(b) Independent and different variances. To account for observations that suggest differing variability for specific landmarks, a model can be constructed that allows for perturbation of each landmark by a different variance. A graphical depiction of such a covariance structure is provided in Figure 3.1b. Differences between Figure 3.1a and 3.1b are apparent.

(c) Correlated landmarks. Since subsets of landmarks can correspond with functionally or developmentally integrated units (Atchley, 1987; Atchley, Cowley et al., 1990; Atchley and Hall, 1991; Atchley, 1993; Cheverud 1995; Cheverud, 1996; Cheverud, Routman et al., 1996), in some cases it may be appropriate to assume correlation among landmark perturbations. For example, if landmark data were available from a long bone, we might expect that the landmarks on the shaft would share certain aspects of variability, while landmarks on a joint surface might share other aspects of variability. Variability of landmarks on female pelves may vary depending on whether the landmark is located at a site associated with locomotion or parturition. A covariance structure can be made somewhat more realistic by allowing correlation among the perturbations local to specific subsets of landmarks. For mathematical convenience, we assume independence between X- and Y-axes. Figure 3.1c depicts this covariance structure.

(d) Correlated landmarks with correlations between the axes.
When we are fortunate in having even more information pertaining to the variation patterns of the landmarks, it may be
possible to add this information to the model. Assume that we
are collecting landmark data on femora of human male adolescent individuals. We know that these teenagers are experiencing
a pronounced growth spurt affecting most parts of the body and
that the magnitude and the exact chronological timing of the
growth spurt vary between individuals. The covariance structure for correlated landmarks (Figure 3.1d) allows us to assume
that landmarks located on the shaft of the femur will have similar covariance structures with correlation between the X- and
Y-axes. In the same model, we can specify an alternate covariance structure for landmarks on the joint surfaces that allows
for a different correlation between the X- and Y-axes.

At this point, we remind the reader of an important caveat for any
statistical or mathematical model. Models are only approximations
that should fit the data as closely as possible. They cannot reflect reality perfectly. This is the reason that the scientist's knowledge of the
data under study should be used when choosing a model. If the statistical model deviates substantially from reality, any statistical analysis
based on that model is of limited utility. This is one reason to strive for
procedures that are valid under a variety of models (Lele, 1991; Lele
and Richtsmeier, 1991).

3.3 Effect of nuisance parameters

Let us return to the transparency experiment where the ten transparencies and the original template are lying in their original
orientation on the table. Now let us suppose that while searching for
the reprint of an important article on covariance structures, the entire
pile of transparencies is knocked to the floor. In the act of retrieving
them from the floor, the red template transparency slips behind a file
cabinet, where it is lost forever. The remaining black pen transparencies are placed back onto the desk and an attempt is made to restore
order to the pile by straightening them (translating and rotating). The
rotation and translation of each transparency away from its original
orientation (the path each transparency took as it flew to the floor, was
retrieved and then straightened within the stack) is unknown and

unknowable. If the path for each transparency were known, the inverse of these paths could be used to reorient them to their original positions. The mean form, M, is lost as well. This mimics the situation encountered by a biologist while collecting data from a sample of biological organisms. The mean form that represents the population average is unknown. Likewise, the random perturbation among specimens, E_i, is unknown and the rotation and translation associated with each form is unknown and unknowable.

Suppose the next day we return to our office and decide to create a few more objects on additional transparencies. We realize that the mean form template (the red transparency) is lost, and we did not record the exact perturbation structure (the covariance matrix of E) that was used when we created the original black triangles. Because our transparencies are circular and of different diameters, there is no particular edge or other outside frame of reference that can be used to put the black transparencies together in a way that may bring us to the original arrangement. We have an informed idea of what the mean form might look like, but no information other than the black transparencies that detail its configuration. This is exactly the situation a scientist faces when examining a set of observations. The scientist has observations but no fundamental frame of reference within which to work. In addition, the mean form and the perturbation structures are unknown.

Questions relating to samples or populations cannot be answered without tools for summarizing the data. The mean form is one summary parameter and the perturbation pattern is yet another. The biological and statistical relevance of the mean form is apparent, while knowledge of the perturbation pattern is the keystone for studying variability. Our primary questions include whether or not we can estimate the mean form from the available observations and whether or not we can estimate the perturbation pattern used to generate the observations.

Estimation of the mean form and the covariance structure in the situation described above is complicated. The complexity comes from the arbitrary and *unknown* parameters of translation and rotation. We have only the displaced black pen transparencies and no knowledge of the original frame of reference (coordinate system) in which they were created. It is up to the investigator to decide if these parameters are critical or even of scientific interest for a specific problem. When studying biological organisms, important biological considerations may enter into this decision. If knowledge of rotation and translation is not germane to the stated scientific problem, it is best to avoid them altogether in analysis. In this context, "avoidance" does not mean inclusion of an arbitrarily chosen rotation and translation system for mathematical convenience. "Avoidance" means that any inferences made from the data should remain the same regardless of, or invariant with respect to, such an arbitrary choice. The existence of translation and rotation affects the estimation of the essential parameters of interest: the mean form M, and the covariance structure Σ. For this reason, translation and rotation are referred to as "nuisance parameters" in statistics and must be considered carefully.

The difficulty of statistical inference in the presence of nuisance parameters is well-studied in statistics. Jerzy Neyman and Elizabeth Scott (Neyman and Scott, 1948) used a simple one-dimensional example to illustrate the effect of nuisance parameters on statistical inference. We consider a similar example here to clarify the concept of nuisance parameter.

Suppose that we have introduced a new drug to treat high blood pressure. We are specifically interested in the intrinsic efficacy of the drug and the variability of the effectiveness. Let Y_{ij} denote *the reduction in blood pressure* for the j-th patient treated in the i-th clinic. Suppose that we treat two patients per clinic and that there are I clinics in the study. Let μ denote the average reduction in blood pressure and σ^2 denote the variability in this effect due to patient variability. Because each clinic also has a slightly different environment, the mean response to the administration of the drug varies from clinic to clinic, but we are not interested in this aspect. Our interest lies in the effect of the drug, μ, and the variability in this effect, σ^2. The model that describes the relationship among the parameters is $Y_{ij} \sim N(\mu + \alpha_i, \sigma^2)$, $i = 1, 2, \ldots, I; j = 1, 2 \ldots$, where α denotes the "clinic effect." Given the research problem and specific research question posed above, the difference in the effect of the drug from clinic to clinic, or the *clinic effect*, α_i, is a "nuisance" parameter. Not only are we not interested in the nuisance parameter, estimation of the nuisance parameter can influence estimation of the parameters of interest.

The field of statistics is based on the idea that as the sample size increases; the amount of information increases and estimates that more closely approach the truth are obtained. An estimator is called a *consistent estimator* if, as sample size increases, the estimator more closely approximates the true value. Nuisance parameters, if not taken into account properly, can lead to statistical anomalies such as nonestimability and inconsistency. Neyman and Scott (1948) showed in the above example that as the number of observations tends to infinity, the maximum likelihood estimate of σ^2 converges to a quantity that is different from the true value of the patient-to-patient variance. They also showed that, in this case, μ is non-estimable. In this example the maximum likelihood estimator of patient-to-patient variability, σ^2, is biased and is also a "statistically inconsistent" estimator.

In Part 2 of this chapter, we prove that certain estimators such as those based on shape coordinates suggested by Bookstein (1991) and Procrustean superimposition estimators suggested by Goodall (1991) are statistically inconsistent. These estimators are inconsistent due to the effects of nuisance parameters (see Lele and Richtsmeier, 1990; Lele, 1991; Lele, 1993).

3.4 Invariance and elimination of nuisance parameters

The existence of nuisance parameters, though not appreciated in many statistical analyses of laboratory or field data, poses significant problems for the estimation of the mean and variance using landmark coordinate data. An inconsistent estimator is unsatisfactory for use in scientific research. Fortunately, there are ways to circumvent these problems.

One approach is to transform the data such that the distribution of the transformed data is independent of the nuisance parameters. Statistical inference is then based only on the transformed data. Consider the blood pressure example introduced earlier. Using the variables introduced in that example we can define a new variable, $W_i = Y_{i1} - Y_{i2}$. The new observation, W_i, represents the difference between observations Y_{i1} and Y_{i2}, and is the difference in the reduction of blood pressure for the two patients at the i-th clinic. This transformation introduces a new model for the distribution of the transformed data. After this transformation, the data W_i's are distributed as $W_i \sim N(0, 2\sigma^2)$. Notice that the distribution of W_i does not depend upon the nuisance parameter α_i, the clinic effect. We have eliminated the nuisance parameter α_i by transforming the data (Y_{i1}, Y_{i2}) to W_i. Removal of the nuisance parameter enables estimation of the variance parameter, σ^2, using the maximum likelihood method (Casella and Berger, 1990) where the likelihood is based on $W_i \sim N(0, 2\sigma^2)$. This estimator is also statistically consistent (Neyman and Scott, 1948).

In the study of landmark coordinate data, nuisance parameters can be eliminated using a similar approach. We first transform the landmark coordinate matrix to what is known as the Euclidean distance matrix that consists of all possible linear distances among the land marks. This transformation eliminates the nuisance parameters of translation, rotation, and reflection. One important consequence of any such transformation adopted for the purpose of eliminating nuisance parameters is that the original parameters of interest, namely, the mean form and the perturbation structure, are transformed as well. As a result, only the transformed parameters are estimable and not the original parameters. But are these *transformed* parameters useful for scientific interpretation? Our experience is that in the case of landmark coordinate data used in biological analysis, the transformed parameters described are meaningful and interpretable.

In the next section, we explain the necessary data transformation and corresponding parameter transformation that eliminate the nuisance parameters of translation, rotation, and reflection. We also demonstrate that the transformed parameters are biologically interpretable.

3.5 A definition of form

To demonstrate that the suggested transformation does not affect the study of form, we must begin with a precise definition of the concept of *form* of an object.

*Definition: The **form** of an object is the characteristic that remains invariant under any translation, rotation or reflection of the object.*

To clarify this definition, consider the simple situation of a triangle, defined by the location of three landmarks. Suppose we rotate or translate or reflect the triangle by an arbitrary amount. Any such movement of the triangle results in changes in the coordinate locations of the three vertices. Although no changes have been made regarding the relative location of the landmarks, a new set of coordinates is required to define the new location of the three landmarks once the triangle has been translated, rotated, or reflected. This means that the landmark coordinate matrix changes upon reflection, translation, or rotation and that the landmark coordinate matrix is not invariant with respect to translation, rotation, and/or reflection.

Now, consider characterizing the form of a triangle as a vector of distances between all possible pairs of landmarks (a vector of three distances in the case of a triangle). This vector of inter-landmark distances is equivalent to the definition of the triangle by landmark locations with one subtle but important difference: a coordinate system is not required to record the inter-landmark distances. While rotation, translation, and reflection change the values of the coordinate locations of landmarks that define the triangle, these operations have no effect on the vector of distances between landmark pairs. A vector of all possible inter-landmark distances representing the relative location of the landmarks is invariant with respect to rotation, reflection, and translation of the triangle.

In fact, a stronger property holds. There is a one to one correspondence between a given triangle and a vector of three distances such that given this vector of distances, one can draw (or reconstruct) the original triangle. For example, in simple Euclidean geometry terminology, two triangles have the same form if they are congruent (Fishback, 1969). When a triangle is drawn from a vector of inter-landmark distances, it will be a triangle *congruent* with the original triangle from which the distances were calculated. Given one type of data, we can construct the other *uniquely* (up to translation, rotation, or reflection). Such a one-to-one, invariant transformation of the data is called a *"maximal invariant."*

At this point, the following statement bears repeating: the landmark coordinate matrix is not invariant to translation, rotation, or reflection, but the corresponding vector of all possible pair-wise distances *is* invariant to these operations. Our example has focused on triangles but these findings are true for objects defined by more than three landmarks in two- or three-dimensional space. Given a landmark coordinate matrix, there is a corresponding and unique vector of all possible pair-wise distances. Conversely, given a vector of all possible distances, one can construct the original landmark coordinate matrix up to translation, rotation, and reflection. We have neither gained nor lost information, whether the form is recorded as a vector of distances or as a landmark coordinate matrix. All the relevant "geometric" information is identical in both representations, and the transformation from one to the other is exact. The main difference between these two representations of the form is that the vector of distances is independent of the choice of an arbitrary orientation, whereas the landmark coordinate matrix is not. The vector of distances is a *coordinate system-free* representation of form. We will show that a vector of all possible distances is biologically interpretable as well.

3.6 Coordinate system-free representation of form

We have described a coordinate system-free representation of a triangle as the Euclidean distances among all possible landmark pairs. This can be easily extended to an object with any number of landmarks, that number represented by K, and any number of dimensions represented by D, such that K is strictly larger than D. Because all the information about the form of an object defined on the basis of landmark coordinates is summarized in the collection of all distances between pairs of landmarks, we call such a collection of distances (put in matrix form) a *form matrix*. The number of unique pair-wise linear distances in a form matrix is L where $L = K(K-1)/2$.

Definition of form matrix: The form matrix or Euclidean distance matrix corresponding to the landmark coordinate matrix A is a matrix consisting of all possible pair-wise distances between landmarks. The form matrix of A is defined as:

$$FM(A) = \begin{bmatrix} 0 & d_{12} & \cdots & \cdots & d_{1K} \\ d_{21} & 0 & \cdots & \cdots & d_{2K} \\ \vdots & \vdots & \vdots & \vdots & \vdots \\ \vdots & \vdots & \vdots & \vdots & \vdots \\ d_{K1} & d_{K2} & \cdots & \cdots & 0 \end{bmatrix}$$

This is a square, symmetric matrix where d_{ij} is the Euclidean distance between landmarks i and j. The form matrix for a given object provides all the relevant information about the form of that object that can be obtained from landmark coordinates.

As an illustration, we present a simple two-dimensional example. We strongly recommend that the reader conduct these calculations and draw the original triangle on a piece of graph paper. The distances between landmarks can be obtained easily using a ruler.

Let the landmark coordinate matrix be $A = \begin{bmatrix} 0 & 0 \\ 1 & 0 \\ 0 & 1 \end{bmatrix}$. This is a triangle with vertices at (0,0), (1,0), and (0,1) for landmarks 1, 2, and 3, respectively. The form matrix corresponding to this triangle is given by the collection of all pair-wise distances in a matrix, namely,

$$FM(A) = \begin{bmatrix} 0 & 1 & 1 \\ 1 & 0 & 1.41 \\ 1 & 1.41 & 0 \end{bmatrix}$$

where columns represent landmarks 1, 2, and 3 read from left to right, and rows represent landmarks 1, 2, and 3 read from top to bottom. The matrix is square because the number of columns is equal to the number of rows. Each cell represents the distance between the landmark that heads a row and the landmark that heads a column. The cell in the upper left-hand corner represents the linear distance between landmark 1 and itself. The second cell in the first row represents the distance between landmarks 1 and 2, while the third cell in the first row represents the distance between landmarks 1 and 3. Notice that the entries on the diagonal are all 0 because the distance between a landmark and itself is zero. The matrix is symmetric, meaning that the entry for the (i,j)-th cell is equal to the entry for the (j,i)-th cell. The first cell in the second row represents the distance between landmarks 2 and 1, which is equal to the value entered in the second cell in the first row, the distance between landmarks 1 and 2.

Since this matrix is symmetric, we need not write it in full matrix format but can abbreviate it by collecting only the above-diagonal elements and writing those numbers as a vector.

$$FM(A) = \begin{bmatrix} 1 \\ 1 \\ 1.41 \end{bmatrix}$$

When written as a vector, the elements to the right of the first diagonal element in row 1 are written in order first, followed by the elements to the right of the diagonal in row 2 and so on. Unless we specify otherwise, the form matrix is always written as a vector.

3.7 The ability to estimate the mean form and variance

In this section, we provide a verbal description of two key results regarding the estimation of the mean form and variance. These results are proven in Part 2 of this chapter. The discussion in this section is targeted at the scientist who is interested in understanding the biological significance of the mathematical results presented in Part 2.

> *Result 1:* The *mean* landmark coordinate matrix, M, consisting of landmark coordinates cannot be estimated. However, the matrix of all possible pair-wise distances corresponding to M, otherwise known as the form matrix of M, or $FM(M)$, *can* be estimated.

An algorithm for the estimation of the mean FM is given in Part 2 of this chapter. If we know the form matrix for a given object, we have all the relevant information about the form of that object that can be obtained from landmark coordinates. This result means that given the landmark coordinate data, we can capture the essence of the mean form by using the vector of all possible linear distances among landmarks. This can be done even in the presence of nuisance parameters of translation, rotation, and reflection.

However, the unfortunate effect of these nuisance parameters becomes apparent when we attempt to estimate the variance. Suppose the variance-covariance matrix V characterizes the perturbation pattern, where $V = \Sigma_K \times \Sigma_D \otimes$ (See Chapter 2, Part 2 for details on Σ_K and Σ_D)

> *Result 2:* Neither Σ_K nor Σ_D can be estimated. What we can estimate is a singular version of Σ_K, denoted by Σ_K^*, and only the eigenvalues of Σ_D.

A consequence of the non-estimability of Σ_K and Σ_D is that interpretation of the estimators of the variances becomes complex. In fact, this constraint means that we cannot estimate the variability local to any particular landmark. However, the eigenvalues of Σ_D can be used to determine whether or not there is a difference in the variability calculated along the three major axes. We underscore that the nuisance parameters prohibit valid estimates of the exact magnitude of variability surrounding landmarks. Consequently, biological questions that require specific values (estimates) of local variability cannot be addressed. Since the estimation of exact quantities of local variability is impossible due to the presence of nuisance parameters, we need to determine how we might use those quantities that can be estimated in scientific analysis.

We refer to the quantities that are estimable, namely Σ_K^* and the eigenvalues of Σ_D, as the perturbation pattern. These estimators can be used as tools to evaluate differences in form or shape between populations, to identify locations or regions that differ least and/or most between two populations, to evaluate a new observation in relation to another group, and to cluster or classify a group of individuals into potential subgroups. The way in which Σ_K^* and the eigenvalues of Σ_D can be used to address many of these issues will be shown in subsequent chapters. Although nuisance parameters limit our inferences many biologically relevant and interesting questions can be addressed using the landmark coordinate matrix data.

    Our conclusion is simple. Landmark coordinate data can be used to obtain a coordinate system-free representation of form, the form matrix, or FM, which is invariant to the operations of translation, rotation and reflection. Landmark coordinate data can also be used to obtain useful features of sample variability. These parameters can be obtained in a statistically sensible and operationally simple fashion. We show in Part 2 of this chapter that these estimators are statistically consistent in that as the sample size increases, the estimators approach the true values. Moreover, these estimators are simple to compute and have high efficiency as compared to other estimators (e.g., maximum likelihood estimators) (Lele and McCulloch, 2000). These topics are more fully developed and studied in Part 2 of this chapter. In the next section, we present estimates of the mean and variance (mean form and variance-covariance matrices) for the biological data sets introduced in Chapter 1.

3.8 Analysis of example data sets

In this section we provide the estimates of the mean form, both in terms of the form matrix (FM) and the landmark coordinate matrix for data sets introduced earlier. Recall that the coordinate matrix representation is useful for pictorially representing the object under study but carries with it all of the problems associated with nuisance parameters. We also present the variance-covariance matrices for these data sets. Throughout this section, we only consider the model where $\Sigma_D = I$. The algorithms used to estimate mean forms and the perturbation covariance structure (Σ_K^*) for each sample are presented in Part 2 of this chapter.

3.8.1 Ts65Dn mouse mandibles

Individuals with Trisomy 21 (Ts21) or Down Syndrome (DS) express different subsets of the phenotypes that characterize the syndrome (e.g., Hirschprungs disease, cardiovascular anomalies, atlanto-axial instability), but the distinct craniofacial appearance seems to occur in all children with DS. Though present in all individuals with DS, the characteristic facial features are highly variable. Since these features most likely arise due to differences in the developmental regime of euploid (non-trisomic) and trisomic individuals, an animal model provides certain advantages to the study of the developmental basis of these facial features since it can be manipulated at any time during development.

The Ts65Dn mouse is a model for DS. To establish the value of the Ts65Dn mouse model in the study of DS, initial energies were directed towards definition of the various phenotypes of interest that are expressed in the animal model. We sought to define the craniofacial phenotype. Using the Reflex microscope, three-dimensional landmark coordinate data were collected from the mandibles of seven adult Ts65Dn mice and 13 adult euploid littermates. Landmarks identify loci on the exterior surface of the left hemi-mandible (see Figure 3.2). These data were used to calculate a mean form, a mean form matrix, and a variance-covariance matrix for both the euploid and aneuploid samples using the methods described above and presented in detail in Part 2 of this chapter. The estimated parameters are given in Table 3.1a-f.

Table 3.1a. *Mean form* of the aneuploid Ts65Dn mandible (*mm*) (Coordinate values correspond to the mean form matrix)

Landmark (number)	X	Y	Z
coronoid process (1)	2.312	2.582	0.408
mandibular angle (2)	5.185	-2.875	-0.044
anterior aspect of condyle (3)	4.702	1.517	-0.070
posterior aspect of condyle (4)	6.231	0.631	-0.460
high point on mandibular body (5)	0.220	-1.872	0.308
posterior aspect at base of alveolar thickening (6)	-1.902	-2.221	-0.113
posterior aspect of incisor at alveolus (7)	-5.192	-0.795	-0.126
anterior aspect of incisor at alveolus (8)	-5.552	1.438	-0.496
body of alveolar ridge (9)	-2.971	-0.269	0.325
anterior aspect of M1 on alveolus (10)	-2.361	0.965	0.039
intersection of coronoid process and alveolar rim (11)	-0.672	0.901	0.229

Table 3.1b. *Mean form* matrix of the euploid Ts65Dn mandible *(mm)*

	1	2	3	4	5	6	7	8	9	10	11
Landmark 1	**0.000**	6.184	2.660	4.446	4.922	6.411	8.246	7.998	6.004	4.958	3.429
Landmark 2	6.184	**0.000**	4.418	3.682	5.077	7.118	10.584	11.580	8.570	8.467	6.973
Landmark 3	2.660	4.418	**0.000**	1.810	5.631	7.589	10.161	10.263	7.888	7.085	5.417
Landmark 4	4.446	3.682	1.810	**0.000**	6.556	8.626	11.517	11.811	9.280	8.613	6.942
Landmark 5	4.922	5.077	5.631	6.556	**0.000**	2.192	5.535	6.702	3.572	3.845	2.914
Landmark 6	6.411	7.118	7.589	8.626	2.192	**0.000**	3.586	5.182	2.268	3.223	3.373
Landmark 7	8.246	10.584	10.161	11.517	5.535	3.586	**0.000**	2.292	2.326	3.338	4.841
Landmark 8	7.998	11.580	10.263	11.811	6.702	5.182	2.292	**0.000**	3.201	3.270	4.963
Landmark 9	6.004	8.570	7.888	9.280	3.572	2.268	2.326	3.201	**0.000**	1.406	2.582
Landmark 10	4.958	8.467	7.085	8.613	3.845	3.223	3.338	3.270	1.406	**0.000**	1.701
Landmark 11	3.429	6.973	5.417	6.942	2.914	3.373	4.841	4.963	2.582	1.701	**0.000**

Table 3.1c. *Variance-covariance* matrix, Σ_K^*, of Ts65Dn mandibles *(mm²)*

	1	2	3	4	5	6	7	8	9	10	11
Landmark 1	**0.03481**	-0.00554	0.00719	0.01942	0.03584	-0.01517	-0.02313	-0.02508	-0.015	-0.00362	-0.00972
Landmark 2	-0.00554	**0.02876**	-0.00745	-0.01084	0.02146	0.01146	-0.0086	-0.01958	-0.00661	-0.00833	0.00527
Landmark 3	0.00719	-0.00745	**0.01558**	0.00719	0.0185	-0.00853	-0.01161	-0.01059	-0.00384	-0.00042	-0.00603
Landmark 4	0.01942	-0.01084	0.00719	**0.02126**	0.02152	-0.01071	-0.01467	-0.0209	-0.00757	0.00197	-0.00667
Landmark 5	0.03584	0.02146	0.0185	0.02152	**0.09834**	-0.0138	-0.05432	-0.07706	-0.02829	-0.0165	-0.00568
Landmark 6	-0.01517	0.01146	-0.00853	-0.01071	-0.0138	**0.01789**	0.00808	0.00665	0.00337	-0.00149	0.00224
Landmark 7	-0.02313	-0.0086	-0.01161	-0.01467	-0.05432	0.00808	**0.03263**	0.04129	0.01583	0.00797	0.00653
Landmark 8	-0.02508	-0.01958	-0.01059	-0.0209	-0.07706	0.00665	0.04129	**0.06808**	0.02286	0.0113	0.00303
Landmark 9	-0.015	-0.00661	-0.00384	-0.00757	-0.02829	0.00337	0.01583	0.02286	**0.01068**	0.00433	0.00424
Landmark 10	-0.00362	-0.00833	-0.00042	0.00197	-0.0165	-0.00149	0.00797	0.0113	0.00433	**0.00733**	-0.00255
Landmark 11	-0.00972	0.00527	-0.00603	-0.00667	-0.00568	0.00224	0.00653	0.00303	0.00424	-0.00255	**0.00935**

Table 3.1d. *Mean form* of the euploid Ts65Dn mandible *(mm)*
(Coordinates corresponding to the mean form matrix)

Landmark (number)	X	Y	Z
coronoid process (1)	3.138	2.903	0.394
mandibular angle (2)	5.414	-3.428	-0.031
anterior aspect of condyle (3)	4.714	1.591	-0.103
posterior aspect of condyle (4)	6.277	0.691	-0.505
high point on mandibular body (5)	0.070	-1.834	0.451
posterior aspect at base of alveolar thickening (6)	-1.932	-2.341	-0.197
posterior aspect of incisor at alveolus (7)	-5.408	-0.845	-0.126
anterior aspect of incisor at alveolus (8)	-5.915	1.492	-0.469
body of alveolar ridge (9)	-3.162	-0.206	0.297
anterior aspect of M1 on alveolus (10)	-2.472	0.999	0.074
intersection of coronoid process and alveolar rim (11)	-0.723	0.978	0.215

There is a direct correspondence between the coordinate data presented for the mean forms and the mean form matrices. Each cell of the
form matrix represents a distance in three-dimensional space that
does not require a coordinate system. For example, the cell that contains the number 6.741 in the mean form matrix of the euploid
mandibles (Table 3.1e) represents the distance between landmarks 1
and 2, a distance that was calculated directly from the landmark coordinate data, but that could be measured on a mouse mandible using
calipers (though perhaps with less precision). The landmark coordinates calculated for the mean form and the mean form matrix are
equivalent expressions; given one, you can derive the other. However,
to express the mean form in terms of landmark coordinates, a coordinate system must be chosen. The mean form matrix does not require
that such a choice be made.

When we compare Σ_K^* for the two samples, the landmarks with
high levels of variation are different for the two samples. Landmarks
3,6,9,10 and 11 show low variance in the euploid sample, while landmarks 10 and 11 appear to be the least variable landmarks in the
aneuploid sample. Also note that the aneuploid Ts65Dn mouse
mandible is more variable overall. Finally, both mandibles show landmarks 5 and 8 to be highly variable.

The variance-covariance matrix provides an indication of variability local to each landmark (variances along the diagonal) and of the
association of variability measures between pairs of points (covariances on the off-diagonals), but nuisance parameters prevent us from

Table 3.1e. *Mean form matrix of the aneuploid Ts65Dn mandible (mm)*

	1	2	3	4	5	6	7	8	9	10	11
Landmark 1	**0.000**	6.741	2.110	3.945	5.644	7.318	9.346	9.202	7.026	5.932	4.318
Landmark 2	6.741	**0.000**	5.068	4.234	5.598	7.428	11.126	12.359	9.167	9.044	7.559
Landmark 3	2.110	5.068	**0.000**	1.848	5.796	7.723	10.411	10.635	8.088	7.212	5.481
Landmark 4	3.945	4.234	1.848	**0.000**	6.769	8.757	11.792	12.219	9.516	8.774	7.044
Landmark 5	5.644	5.598	5.796	6.76	**0.000**	2.165	5.596	6.908	3.622	3.824	2.930
Landmark 6	7.318	7.428	7.723	8.757	2.165	**0.000**	3.785	5.534	2.513	3.394	3.556
Landmark 7	9.346	11.126	10.411	11.792	5.596	3.785	**0.000**	2.416	2.373	3.473	5.038
Landmark 8	9.202	12.359	10.635	12.219	6.908	5.534	2.416	**0.000**	3.324	3.521	5.262
Landmark 9	7.026	9.167	8.088	9.516	3.622	2.513	2.373	3.324	**0.000**	1.407	2.712
Landmark 10	5.932	9.044	7.212	8.774	3.824	3.394	3.473	3.521	1.407	**0.000**	1.754
Landmark 11	4.318	7.559	5.481	7.044	2.930	3.556	5.038	5.262	2.712	1.754	**0.000**

Table 3.1f. *Variance-covariance matrix Σ_K^* of euploid Ts65Dn mandible (mm^2)*

	1	2	3	4	5	6	7	8	9	10	11
1	**0.01796**	0.00512	0.00718	0.00821	0.00862	-0.00567	-0.01313	-0.01721	-0.01001	-0.00305	0.00198
2	0.00512	**0.01282**	0.00179	0.00047	-0.00674	0.00032	-0.00234	-0.00654	-0.00525	-0.0031	0.00346
3	0.00718	0.00179	**0.00744**	0.00866	-0.00018	-0.00375	-0.00658	-0.01071	-0.00578	0.00025	0.00168
4	0.00821	0.00047	0.00866	**0.02336**	-0.01071	-0.00821	-0.00698	-0.01029	-0.00608	0.00128	0.00029
5	0.00862	-0.00674	-0.00018	-0.01071	**0.03549**	-0.00418	-0.01215	-0.00808	-0.00448	-0.00233	0.00473
6	-0.00567	0.00032	-0.00375	-0.00821	-0.00418	**0.00876**	0.00395	0.00412	0.00542	0.00036	-0.00113
7	-0.01313	-0.00234	-0.00658	-0.00698	-0.01215	0.00395	**0.01623**	0.01692	0.00839	0.0002	-0.00453
8	-0.01721	-0.00654	-0.01071	-0.01029	-0.00808	0.00412	0.01692	**0.02379**	0.01163	0.00276	-0.0064
9	-0.01001	-0.00525	-0.00578	-0.00608	-0.00448	0.00542	0.00839	0.01163	**0.00822**	0.0011	-0.00314
10	-0.00305	-0.0031	0.00025	0.00128	-0.00233	0.00036	0.0002	0.00276	0.0011	**0.00557**	-0.00305
11	0.00198	0.00346	0.00168	0.00029	0.00473	-0.00113	-0.00453	-0.0064	-0.00314	-0.00305	**0.00611**

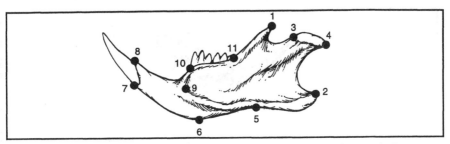

Figure 3.2. Left mouse hemi-mandible showing 11 landmarks used in analysis.

obtaining a valid estimate of the exact magnitudes of variability local to each landmark (see Result 2, page 77). Whether a measure of co-variability is positive or negative is not meaningful in our use of the variance-covariance matrices. Even though we cannot treat the diagonal values from the variance-covariance matrix as proper estimates of variability values, our estimates of the mean form and the variance-covariance matrix can be used to obtain a visual representation of how the location of landmarks varies across a sample of specimens. We do this by generating hypothetical forms using parameters computed from the original data. The resulting data can be used to provide a visual impression of how variability is distributed across the sample of specimens.

As an example, the mean form matrix and variance-covariance matrix for the normal and Ts65Dn mandibles were used to generate hypothetical populations of 50 specimens for each sample (Figure 3.3). To generate data like those represented in Figure 3.3, a mean form matrix is estimated following algorithm 1 (Chapter 3, Part 2) and is transformed back to landmark coordinates using an algorithm 1. The landmark coordinates of the mean form are used as a template, just as the original transparency was used to generate triangles in our pre-ceding exercise with transparencies. However, instead of randomly placing points around each of the given landmarks as we did with the transparencies, the points are placed according to the Gaussian distri-bution with the estimated variance-covariance matrix. The hypothetical data displayed in Figure 3.3 (Richtsmeier et al., 2000) are constrained by the variances calculated local to each landmark and by the covariances that represent the relationships among variances for all landmarks. The diagram drawn of the hypothetical data (Figure 3.3) enables visualization of differences in variances local to specific landmarks within a sample. The true sample size, of course, remains small. Remember that the choice of coordinate system into which to

Figure 3.3. Fifty Gaussian random observations generated for the sample of Ts65Dn aneuploid mice (left) and the sample of Ts65Dn normal littermates (right). The random observations for each sample were generated using the mean form and variance-covariance matrix estimated from the landmark coordinate data (after Richtsmeier et al., 2000). This graphical depiction of the variability for each sample is not invariant with respect to the coordinate system as we are viewing the three-dimensional data from a given perspective. A mouse hemi-mandible is shown to indicate approximate location of landmarks on the mean form.

project the mean form is arbitrary, and that the visual impression of the simulated data can change with the coordinate system. Generation of the hypothetical data is simply a tool to help visualize variability local to landmarks.

It is important to understand that within any sample, estimation of $\Sigma_K{}^*$ changes depending upon the landmarks included in its estimation. Suppose that instead of analyzing all eleven landmarks on the aneuploid Ts65Dn mouse mandible, we decide to analyze a subset of the original group of landmarks, say landmarks 1, 2, 3, 5, 7, and 10. The estimate of the parameters of interest using only those data for the landmark subset are given in Tables 3.2a-c.

Table 3.2a. *Mean form* of the aneuploid Ts65Dn mandibles using the land-mark subset (Coordinates corresponding to the mean form matrix)

Landmark (number)	X	Y	Z
Landmark 1	1.504	2.670	-0.269
Landmark 2	4.370	-2.801	-0.009
Landmark 3	3.891	1.588	0.286
Landmark 5	-0.588	-1.783	-0.152
Landmark 7	-6.005	-0.718	0.104
Landmark 10	-3.172	1.044	0.041

Table 3.2b. *Mean form* matrix of the aneuploid Ts65Dn mandible using the landmark subset *(mm)*

	1	2	3	5	7	10
Landmark 1	**0.000**	6.182	2.678	4.922	8.247	4.961
Landmark 2	6.182	**0.000**	4.425	5.064	10.583	8.467
Landmark 3	2.678	4.425	**0.000**	5.623	10.162	7.088
Landmark 5	4.922	5.064	5.623	**0.000**	5.527	3.835
Landmark 7	8.247	10.583	10.162	5.527	**0.000**	3.336
Landmark 10	4.961	8.467	7.088	3.835	3.336	**0.000**

Table 3.2c. *Variance-covariance* matrix $\Sigma_K{}^*$ of the aneuploid Ts65Dn mandibles using the landmark subset (mm^2)

	1	2	3	5	7	10
1	**0.01633**	-0.01579	0.00444	0.02117	-0.02009	-0.00606
2	-0.01579	**0.02376**	-0.00804	0.00017	0.00428	-0.00438
3	0.00444	-0.00804	**0.00286**	-0.00329	0.00111	0.00292
5	0.02117	0.00017	-0.00329	**0.07947**	-0.06613	-0.0314
7	-0.02009	0.00428	0.00111	-0.06613	**0.05744**	0.02339
10	-0.00606	-0.00438	0.00292	-0.0314	0.02339	**0.01553**

When we compare these estimates with those based on all eleven land-marks (Tables 3.1a to 3.1c), we notice that corresponding elements of the mean form expressed as a form matrix are identical (within numerical accuracy) to those of the form matrix estimated using the full landmark set. However, landmark coordinate locations in X,Y,Z space are quite different between the estimates. This occurs because placement of the mean form into a coordinate space is arbitrary. Elements of $\Sigma_K{}^*$ calculated from the original data and the landmark subset are also different. The reason for this difference is that estimates reported by $\Sigma_K{}^*$ are dependent upon the centering that is done during estimation (see Part 2 of this chapter). Estimation of the mean form matrix does not require centering, but estimation of the variance-covariance matrix does. If a different set of landmarks is used, a different centering occurs, and the variance-covariance matrix changes. Importantly, values reported in the form matrix remain the same whether the entire set of landmarks or a subset are used in the calculation. The mean form matrix is coordinate system-free, but $\Sigma_K{}^*$ is not. For this reason, specific values of the variance-covariance matrix should not be directly compared between samples. Relative magnitudes of variance estimates within an estimated variance-covariance matrix can be compared.

The samples presented thus far were chosen as examples of relatively small sample size. Small sample sizes are common in biological research, and the investigator needs to be aware of the consequences for the estimation of parameters and for the statistical testing of differences in forms. There are often economic, scientific, or biological reasons for small sample sizes in research situations. In our example, the aneuploid Ts65Dn sample is small because the transmission of segmental trisomy to offspring is well below 100%, and of the segmentally trisomic animals produced in any litter, only the females are fertile. Moreover, the mice are expensive to house and breed. When faced with a small sample size, our experience suggests that estimates based on method of moments techniques may be more stable than those based on maximum likelihood techniques (Lele and McCulloch, 2000). It should be kept in mind, however, that any estimator might fail given small sample sizes.

Small sample sizes can impact the estimation of certain parameters. When calculating the mean form matrix, a quantity is calculated (see Algorithm 1, Step 5, Part 2 of this chapter) that represents the difference between the sample mean of the squared Euclidean distance between two landmarks and the sample variance of the squared Euclidean distance. When the variance of the squared distance is larger than the square of the mean distance between two landmarks, this quantity is negative, which causes obvious problems. This situation is more likely to occur in small samples among landmarks that are closely spaced and is less likely to occur as sample size increases.

The most effective way to avoid this problem during estimation is

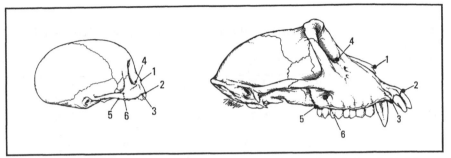

Figure 3.4. Facial landmarks plotted on the juvenile (left) and adult female (right) skull of *Macaca fascicularis* (scale is approximate).

by increasing the sample size. When increasing the sample size is not possible, we suggest the following alternate solution. Observe the landmarks that lie in close proximity to one another on the form, and decide on the basis of the scientific question under consideration and your knowledge of the forms whether both landmarks are truly needed for analysis. If both landmarks are important, the data can be analyzed initially with only one of the two neighboring landmarks, and then analyzed again with the other.

3.8.2 Crab-eating macaque facial skeleton

Three-dimensional landmark data from a sample of juvenile (N= 26) and adult (N=30) male crab-eating macaque (*Macaca fascicularis*) skulls provide the next example. Individual skulls were aged by observing which teeth were present (Richtsmeier and Cheverud et al., 1993). Developmental age groups were formed based on previous studies correlating tooth eruption with chronological ages in laboratory specimens. The juvenile developmental age data set consists of individuals who had any combination of their deciduous teeth and first permanent molar erupted. The adult developmental age data set consists of individuals who have all permanent teeth in full eruption.

Data representing six landmarks on the right side of the facial skeleton (Figure 3.4) of the crab-eating macaques were collected using the 3Space tabletop digitizer (units are in *cm*). From these data, a mean form matrix, a mean form, and a variance-covariance matrix were calculated for each age group. These data are given in Table 3.3a-f.

Table 3.3a. *Mean form* matrix of the juvenile *Macaca fascicularis* facial skeleton (*cm*)

	1	2	3	4	5	6
1	**0.000**	1.273	1.546	1.161	2.574	2.334
2	1.273	**0.000**	0.929	1.739	2.558	2.520
3	1.546	0.929	**0.000**	1.283	1.994	2.237
4	1.161	1.739	1.283	**0.000**	1.857	1.937
5	2.574	2.558	1.994	1.857	**0.000**	0.842
6	2.334	2.520	2.237	1.937	0.842	**0.000**

Table 3.3b. *Mean form* of the juvenile *Macaca fascicularis* facial skeleton (*cm*) (Coordinates corresponding to the mean form matrix)

Landmark (number)	X	Y	Z
Nasale (1)	0.918	0.726	-0.266
Intradentale superior (2)	1.070	-0.529	-0.420
Premaxilla-maxillary junction (3)	0.556	-0.648	0.345
Zygomaxillare superior (4)	0.199	0.551	0.628
Maxillary tuberosity (5)	-1.400	-0.292	0.199
Posterior nasal spine (6)	-1.343	0.192	-0.488

Table 3.3c. *Variance-covariance* matrix, Σ_K^*, of the juvenile *Macaca fascicularis* facial skeleton (*cm²*)

	1	2	3	4	5	6
1	**0.01858**	-0.00085	0.0062	0.00573	-0.0191	-0.01056
2	-0.00085	**0.04619**	0.02962	-0.02073	-0.02768	-0.02655
3	0.0062	0.02962	**0.03353**	-0.02394	-0.02752	-0.01789
4	0.00573	-0.02073	-0.02394	**0.0245**	0.00816	0.00627
5	-0.0191	-0.02768	-0.02752	0.00816	**0.04132**	0.0248
6	-0.01056	-0.02655	-0.01789	0.00627	0.0248	**0.02393**

Table 3.3d. *Mean form* matrix of adult *Macaca fascicularis* facial skeleton (*cm*)

	1	2	3	4	5	6
1	**0.000**	2.357	2.239	1.695	3.741	3.104
2	2.357	**0.000**	1.214	3.475	4.387	4.196
3	2.239	1.214	**0.000**	2.740	3.621	3.672
4	1.695	3.475	2.740	**0.000**	2.818	2.447
5	3.741	4.387	3.621	2.818	**0.000**	1.249
6	3.104	4.196	3.672	2.447	1.249	**0.000**

Table 3.3e. *Mean form* of the adult *Macaca fascicularis* facial skeleton (*cm*) (Coordinates corresponding to the mean form matrix)

Landmark (number)	X	Y	Z
Nasale (1)	0.860	-1.194	0.455
Intradentale superior (2)	2.192	0.750	0.392
Premaxilla-maxillary junction (3)	1.461	0.700	-0.577
Zygomaxillare superior (4)	-0.424	-1.288	-0.648
Maxillary tuberosity (5)	-2.141	0.917	-0.279
Posterior nasal spine (6)	-1.947	0.114	0.657

Table 3.3f. *Variance-covariance* matrix, Σ_K^*, of the adult *Macaca fascicularis* facial skeleton (*cm²*)

	1	2	3	4	5	6
1	**0.0164**	-0.00109	0.00028	-0.0034	-0.00747	-0.00471
2	-0.00109	**0.01942**	0.00775	-0.01604	-0.00442	-0.00561
3	0.00028	0.00775	**0.00833**	-0.00654	-0.00598	-0.00383
4	-0.0034	-0.01604	-0.00654	**0.01605**	0.00356	0.00637
5	-0.00747	-0.00442	-0.00598	0.00356	**0.01176**	0.00255
6	-0.00471	-0.00561	-0.00383	0.00637	0.00255	**0.00524**

Notice that variances calculated for the landmarks in the adult face are relatively small. Some of this is scale related; variances are smaller with respect to the distances between the landmarks in larger specimens. In addition, the samples are defined on the basis of developmental ages, based on tooth eruption patterns. The adult sample contains individuals who have a complete set of permanent teeth. The sample of immature individuals on the other hand includes skulls of individuals who had very few deciduous teeth erupted, as well as individuals who have all deciduous dentition erupted including the first molar. The juvenile sample therefore contains individuals at various stages during a particularly intense period of growth.

Two landmarks, intersection of the premaxilla and maxilla on the alveolus (landmark 3) and posterior nasal spine (landmark 6), display variances that are smaller than the variances calculated for the other points in the adult sample (see the 3rd and 6th entries along the diagonal for Σ_K^* for the adults, Table 3.3f). The premaxillary-maxillary intersection marks the most anterior point for the canine. The posterior nasal spine is the most posterior point on the hard palate, defining the length of the hard palate. Biological constraints concerning uniformity in dentition, requirements of the palate to separate the oral and nasal orifices, and attachment sites for soft tissues, may account for the relatively reduced variances for these two landmarks. The interpretations that we offer as potential reasons for local differences in variation come from prior knowledge of the anatomy and development of the primate skull, and should be further investigated. Prior knowledge from any relevant discipline can and should be used to interpret local differences in variation.

3.8.3 Neurocranial morphology in craniosynostosis

Eight landmarks from the external surface of the neurocranium (calotte or skullcap) were collected from computed tomography (CT) scans of 16 infants and children diagnosed with premature closure of the sagittal suture. These children range in age from six weeks to seven years of age. The mean form matrix, mean form, and variance-covariance matrix for these data are given in Table 3.4a-c.

Table 3.4a. *Mean form* of the calotte of children with sagittal synostosis (*mm*). (Coordinates corresponding to the mean form matrix).

Landmark (number)	X	Y	Z
Right asterion (1)	41.509	42.599	24.623
Left asterion (2)	43.008	-40.134	26.122
Bregma (3)	-38.021	-1.320	-40.743
Lambda (4)	68.148	-0.770	-27.897
Right frontal boss (5)	-68.448	27.304	12.088
Left frontal boss (6)	-67.125	-28.598	13.242
Right parietal boss (7)	9.464	50.580	-3.954
Left parietal boss (8)	11.464	-49.661	-3.481

Variances are very large in this small sample. One reason for this lies in the extremely large age range represented by the children that make up this sample. Given that more than two thirds of adult head size is attained by two and one half years of age, this sample includes an extreme range of sizes, from specimens very close to the smallest human postnatal size possible to those approaching 75% of their eventual adult head size. Additionally, the frontal and parietal bosses (landmarks 5, 6, 7, 8) represent "fuzzy landmarks" (see Chapter 2), the measurement error of which has been shown to be relatively larger than that of more traditional landmarks (Valeri et al., 1998).

Table 3.4b. *Mean form matrix for the sagittal synostosis sample (mm)*

	1	2	3	4	5	6	7	8
Landmark_1	**0.000**	82.760	111.923	73.136	111.721	130.384	43.672	101.018
Landmark_2	82.760	**0.000**	111.997	71.412	131.024	111.483	101.286	44.297
Landmark_3	111.923	111.997	**0.000**	106.945	67.351	67.123	79.384	78.576
Landmark_4	73.136	71.412	106.945	**0.000**	145.070	144.103	81.571	78.737
Landmark_5	111.721	131.024	67.351	145.070	**0.000**	55.929	82.882	112.036
Landmark_6	130.384	111.483	67.123	144.103	55.929	**0.000**	111.494	83.064
Landmark_7	43.672	101.286	79.384	81.571	82.882	111.494	**0.000**	100.263
Landmark_8	101.018	44.297	78.576	78.737	112.036	83.064	100.263	**0.000**

Table 3.4c. *Variance-covariance matrix, Σ_K^*, of the sagittal synostosis sample (mm^2)*

	1	2	3	4	5	6	7	8
1	126.2355	27.65825	-63.3063	33.3013	-57.8084	-63.69	32.42553	-34.8159
2	27.65825	112.6592	-61.2221	33.54741	-57.5014	-52.5317	-40.4518	37.84221
3	-63.3063	-61.2221	59.42549	-44.2029	59.33071	58.88485	0.52835	-9.43802
4	33.3013	33.54741	-44.2029	78.37691	-90.3996	-88.2266	33.47435	44.12919
5	-57.8084	-57.5014	59.33071	-90.3996	123.0724	119.9785	-43.4827	-53.1895
6	-63.69	-52.5317	58.88485	-88.2266	119.9785	119.5009	-44.6665	-49.2495
7	32.42553	-40.4518	0.52835	33.47435	-43.4827	-44.6665	61.69917	0.47355
8	-34.8159	37.84221	-9.43802	44.12919	-53.1895	-49.2495	0.47355	64.24797

Variances calculated for right and left asterion (sutural intersections) and for the frontal bosses (fuzzy landmarks) are similar and relatively large. The variance for bregma (3), lambda (4), and the parietal bosses (7 and 8) (fuzzy landmarks) are similar to one another and relatively small. This example demonstrates that the increased measurement error in locating fuzzy landmarks (Valeri et al., 1998) may not always result in fuzzy landmarks displaying greater measures of variance relative to other landmark types.

Markedly high variances for any landmark is a condition that needs to be closely studied in any morphometric analysis, especially when landmarks with large variances are located close to one another on a form. When EDMA is applied and the model-based bootstrapping method is chosen for statistical testing, hypothetical forms will be generated from the mean form and variance-covariance matrix. When the variability around a landmark is large and landmarks are placed relatively close to one another (Figure 3.5), it is possible that two landmarks can change relative positions within a form created for the testing procedure. The data shown in Figure 3.5 could produce a simulated form with landmarks swapping relative locations, resulting in an error that would prohibit statistical testing.

3.9 Some comments on EDMA
vs. other morphometric methods

Reflection Invariance. Earlier we defined form as "that characteristic which remains invariant under translation, rotation, or reflection of the object." In some of the approaches to the analysis of landmark coordinate data, reflection invariance is not included in the definition of form (e.g., Goodall, 1991). However inclusion of reflection may substantially simplify statistical and mathematical analysis and may be advantageous in certain research situations. Reflection invariance does not compromise biological understanding or intuition, because any scientist who has collected data knows those data, as well as the data he or she would ideally like to have for analysis. Consider the following example. A paleontologist is trying to understand the evolution of a species from fossilized teeth. She is collecting landmark coordinate data from cusp tips and other landmarks identified on the occlusal surface of teeth. Different tooth types (incisors, canines, premolars, and molars) cannot be combined to increase the sample size for certain poorly represented species. However, if in some individuals only a right

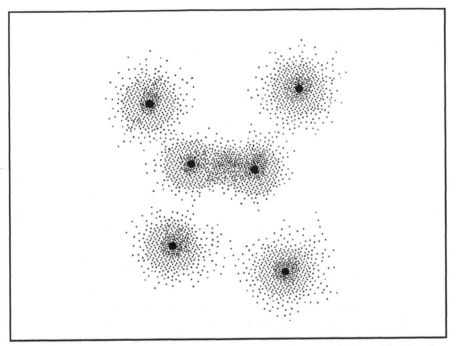

Figure 3.5. A hypothetical example of landmark data whose proximity and local variability cause landmarks to swap relative locations during statistical analysis prohibiting statistical testing. When EDMA is applied and the model-based bootstrapping method is chosen for statistical testing (see Chapter 4), hypothetical forms are generated from the mean form and variance-covariance matrix. When the variability around a landmark is large and landmarks are placed relatively close to one another, as in the center of this figure, two landmarks can change relative positions preventing statistical testing.

first molar is available, while in other individuals only a left first molar is available, these can be combined into a single sample if symmetry of the biological organisms can be assumed and if the methods used are reflection invariant. Of course the biologist will keep records of the collected data and have knowledge of which specimens were reflected for analysis. The concept of reflection invariance can be useful mathematically, statistically, and biologically.

Nuisance parameters and variance estimation. We have proposed a method for estimating the mean form and the variability of forms within a single population using the Euclidean Distance Matrix representation of the form of an object. An alternate method for the estimation of the mean form and variability of forms within a single population is based on Procrustes superimposition (Goodall, 1991).

Lele (Lele, 1993) provided the first evidence of errors associated with this intuitively appealing approach, which were subsequently supported and extended by Kent and Mardia (1997). The problems associated with the Procrustes method when applied under the Gaussian perturbation model used in this chapter (and used by Goodall, 1991) have consequences for data analysis. Choosing a single coordinate system for Procrustes superimposition does not *eliminate* the nuisance parameters. Instead, it constrains the nuisance parameters to take a certain form. Moreover, because the nuisance parameters are not eliminated properly, the Procrustes mean form and mean shape estimators are statistically inconsistent. This means that as the sample size increases, the estimator converges to a quantity that is different from the true mean form, or the true mean shape. Even more important, however, is the observation that the variance-covariance estimator is statistically inconsistent. We have found that even when the amount of error in the estimation of the mean form or mean shape is small, the error in the covariance estimator can be substantial.

The implications of inconsistency of the variance-covariance estimator are serious. Any statistical inference procedure that uses the Procrustes estimator of variance will yield incorrect results. Confidence intervals for form or shape difference cannot have correct coverage probabilities if the variance estimators are wrong. For example, when the variance estimators are wrong and it is claimed that we have a 95% confidence interval, the true value may not actually be covered 95% of the time. Similarly, Principal Components Analysis based on the Procrustes residuals (Kent, 1994) can be patently misleading when variance is estimated incorrectly. In biology, perhaps more than in any other science, variability is one of the most important parameters than can be estimated using statistics. The Procrustes method fails in its estimation of this parameter.

The following example, which our readers can carry out using transparencies, illustrates how the Procrustes approach fails to correctly measure variability (a rigorous mathematical discussion is provided in Part 2 of this chapter). Let us begin by examining how the Procrustes superimposition is implemented. Consider the two quadrangles in Figure 3.6. On one transparency, use a red pen to draw the solid quadrangle, and on another transparency use a green pen to draw the dashed quadrangle. The centroid of each quadrangle is defined as the point at which the two diagonals cross. The first step in any superimposition method is to fix one of the figures (say the red quadrangle) and translate the other figure (the green quadrangle) so that it matches the first figure at some point. In the Procrustes superimposition, the first step is to translate the second figure so that the centroids of the two figures match. Once this is done, no further translation is allowed and the relative position is held constant during the rotation step. Mimic the translation step by overlaying the transparencies to match the centroids and push a pin through the superimposed centroids. Now rotate

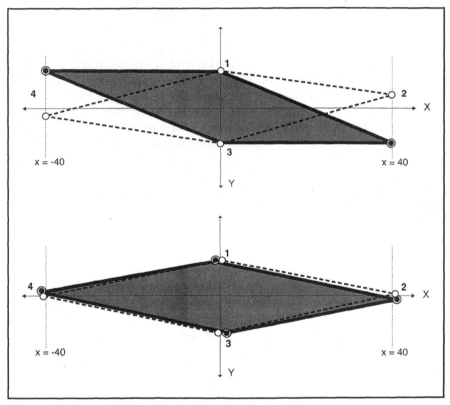

Figure 3.6. Two quadrilaterals, each defined by four, two-dimensional landmarks superimposed on their centroids. This superimposition demonstrates a basic tendency of the Procrustes fitting criterion: corresponding landmarks farthest from the centroid are matched closely at the cost of mismatching those that are closer to the centroid. Additionally, after rotation, the landmark locations of points 2 and 4 over-estimate the distance between landmarks 2 and 4. Similarly, the placement of landmarks 1 and 3 after rotation tend to underestimate the distance between landmarks 1 and 3 along the Y-axis (not readily apparent in this figure; after Lele, 1993). Moreover, estimates of variability for those landmarks lying farther from the centroid are reduced, while estimates of variability for landmarks lying close to the centroid are amplified (data not shown).

the green quadrangle in an attempt to minimize the sum of the squared distances between corresponding landmarks in the two quadrangles.

Note that if we disregard landmarks 1 and 3, a rather nice fit is accomplished between landmarks 2 and 4 of the red and green transparencies. As the green quadrangle is rotated to match landmarks 1 and 3 of the green and red quadrangles, we tend to lose the fit between landmarks 2 and 4. On the other hand, as landmarks 2 and 4 align closely, landmarks 1 and 3 are not well matched.

This simple example demonstrates a basic tendency of the Procrustes fitting criterion: corresponding landmarks farthest from the centroid are matched closely at the cost of mismatching those that are closer to the centroid. A consequence is that estimates of variability for those landmarks lying farther from the centroid are reduced, while estimates of variability for landmarks lying close to the centroid are amplified. This effect will be most pronounced when the object is not symmetric around the centroid and when the landmarks are not uniformly spread on the object. This means that the Procrustes method tends to estimate variability according to a rule that has little to do with the natural variability of the specimens, but is instead driven by the distance of landmarks from the centroid.

3.10 Summary

In Part 1 of this chapter, we provided the reader with working models for landmark data and introduced concepts that should be considered when analyzing landmark data. The terms *model* and *method* were defined and related to one another. We defined *nuisance parameters* as they exist in the study of form. We introduced the concept of *invariance* and how it relates to elimination of the nuisance parameters, and presented a *coordinate system-free* representation of form. We presented statistical models for landmark data and explained the estimators for the one-sample case using Euclidean Distance Matrix Analysis. Finally, we commented upon aspects of our approach to the analysis of landmark data that were not essential to the central presentation of our ideas, but that place our approach within the context of the field of morphometrics. Part 2 of this chapter deals with the mathematical and statistical details of the one-sample case.

Statistical Theory for the Analysis of a Single Population

In this part, we discuss the statistical theory involved in the analysis of data obtained from a single population. In particular, we describe the use of invariance to eliminate nuisance parameters. The distribution of the maximal invariant is used to shed some light on the identifiability of the parameters of interest. The method of moments is used to obtain estimators of the parameters of interest. It is argued that these estimators are consistent and are computationally simple to obtain. We also compare and contrast competing methods of estimation: the Procrustes method, the shape coordinates method, and the method of maximum likelihood from both the theoretical point of view and the practical point of view. This discussion is based on Lele (1993) and Lele and McCulloch (2000).

3.11 The perturbation model

We model the inter-individual variability by the Gaussian perturbation model that has been previously described (Goodall, 1991 or Lele, 1993). The perturbation model may be thought of as representing the following process. To generate a random geometrical object or equivalently a K point configuration in D dimensional Euclidean space, one must first choose a mean form (represented by matrix M) and perturbs the elements of this matrix by adding noise to this mean form according to a matrix valued Gaussian distribution. The K point configuration so obtained is then rotated and/or reflected by an unknown angle and translated by an unknown amount. Such perturbed, translated, rotated, and/or reflected K point configurations constitute our data. The above description can be put in a mathematical form as follows.

Let M be a $K \times D$ matrix corresponding to the mean form. Let E_i be a $K \times D$ matrix-valued Gaussian random variable (Arnold, 1981, pages 309-323) representing the error for the i-th individual. We assume E_i to be Gaussian with a mean matrix of 0 and covariance matrices given by Σ_K and Σ_D. The matrix Σ_K describes the covariances between elements within the same column of E_i and Σ_D describes the covariances within the rows of E_i. In terms of notation we have $E_i \sim N(0, \Sigma_K, \Sigma_D)$. More precisely, if we stack the matrix, E_i, into a vector, $vec(E_i)$, then we have $var(vec(E^T)) = \Sigma_K \otimes \Sigma_D$. Let R_i be an orthogonal matrix corresponding to the rotation of the i-th individual, and let t_i be a $1 \times D$ matrix corresponding to the translation of the i-th individual. Then the landmark coordinate matrix corresponding to the i-th individual may be represented as: $X_i = (M + E_i)R_i + \underline{1}t_i$ where $\underline{1}$ is a $K \times 1$ matrix of 1's. The random matrices X_i thus follow: $X_i \sim N(MR_i + \underline{1}t_i, \Sigma_K, R_i\Sigma_D R_i^T)$.

An important thing to notice is that the matrices R_i and t_i are unknown and unknowable. These are nuisance parameters, while the parameters of interest are (M, Σ_K, Σ_D). In fact, there are more unknown parameters than there are observations and the number of parameters grows with the sample size. This problem falls within the class described by Neyman and Scott (1948).

3.12 Invariance and elimination of nuisance parameters

One way to eliminate the nuisance parameters is by considering a maximal invariant under a group of transformations. We briefly review the definition of a maximal invariant and an important consequence of using a maximal invariant for the statistical inference to the identifiability of the underlying parameters. For an excellent discussion of invariance, we refer the reader to Lehmann (1959, Chapter 6), Berger (1982, Chapter 6) or Arnold (1981, Chapter 1).

Definition 1: Let G be a group of invertible functions from a set C to itself. A function $T(c)$ is called a maximal invariant if it satisfies the following two conditions:

a) $T(g(c)) = T(c)$ for all $g \in G$ and $c \in C$.
b) If $T(c_1) = T(c_2)$, then there exists $g \in G$ such that $c_2 = g(c_1)$.

For any set C and any group G of invertible functions from C into itself, there exists a maximal invariant. Any one to one function of a maximal invariant is itself a maximal invariant. Hence, maximal invariants are not unique. However, all maximal invariants are equivalent in the sense that their sets of constancy coincide.

Let X be distributed according to a probability distribution P_θ, $\theta \in \Theta$. Denote by gX the random variable that takes on the value gx when $X=x$, and suppose that when the distribution of X is P_θ, the distribution of gX is $P_{\theta'}$, with θ' also in Θ. The element θ' associated with θ in this manner will be denoted by $\bar{g}\theta$. The transformation $\bar{g}$ of Θ onto itself, defined in this manner, is one to one provided the distributions P_θ corresponding to different values of θ are distinct. The action of the group $\bar{G}$ induces a partition of the parameter space Θ into equivalency classes.

Invariance, by reducing the data to a maximal invariant statistic T, typically also shrinks the parameter space. Let $T(x)$ be a maximal invariant under G and let the distribution of $T(x)$ depend on $v(\theta)$. Then the inverse image of $v(\theta)$ partitions the original parameter space into equivalency classes. All the points that belong to the same equivalency class map onto $v(\theta)$ and conversely, if two parameter values θ_1 and θ_2 are such that $v(\theta_1)=v(\theta_2)$, then θ_1 and θ_2 belong to the same equivalency class. The partition of Θ under the group $\bar{G}$ and the partition induced by $v(\theta)$ do not necessarily match each other.

We illustrate these ideas using a simple example involving only the translation group. Let X_i be a bivariate random variable and let $X_i \sim N(\mu,\Sigma)$. Then the parameter space Θ is given by $\Theta = \{(\mu,\Sigma): \mu \in R^2$ and Σ is any 2 x 2 real, symmetric, positive definite matrix$\}$. Let the group action consist of translations only. That is, $gX_i = X_i + \underline{1}t_i$. It then follows that $gX_i \sim N(\mu+\underline{1}t_i,\Sigma)$ where $\underline{1}$ is a 2 x 1 vector of 1's, t_i is a real number and Σ is a 2 x 2 real, symmetric, positive definite matrix. The orbit corresponding to any given (μ^*,Σ^*) under the group $\bar{G}$ is given by $\{(\mu,\Sigma)\in\Theta:\mu=\mu+\underline{1}t,\Sigma=\Sigma^*\}$.

A maximal invariant under the translation group in the above situation is given by: $Y_i = (X_{i2}-X_{i1})$. The distribution of this maximal invariant is $N(\mu_2-\mu_1,\sigma_{11}+\sigma_{22}-2\sigma_{12})$. Let us denote $\mu_2-\mu_1=\delta_\mu$ and $\sigma_{11}+\sigma_{22}-2\sigma_{12}=\phi$. Then, given the values of (δ_μ,ϕ), the inverse image in the original parameter space Θ is given by $\{(\mu,\Sigma)\in\Theta:\mu_2-\mu_1=\delta_\mu,\sigma_{11}+\sigma_{22}-2\sigma_{12}=\phi\}$. This defines a partition of Θ which is different than the partition defined by the group $\bar{G}$. Notice that the partition defined by the maximal invariant $Y_i=(X_{i2}-X_{i1})$ is more coarse than the partition defined by $\bar{G}$. If invariance is used, only the partitions defined by $v(\theta)$ are identifiable. This is the cost we pay due to the presence of nuisance parameters.

Now let us consider the landmark coordinate data problem. The landmark coordinate matrix is denoted by X. Recall that the group of transformations involved in the analysis of landmark coordinates consists of translation, rotation and possibly reflection. Thus $gX = XR + \underline{1}t$. It is easy to see that if $X \stackrel{d}{=} N(M, \Sigma_K, \Sigma_D)$, then $gX \stackrel{d}{=} N(MR + \underline{1}t, \Sigma_K, R^T \Sigma_D R)$, where $\stackrel{d}{=}$ means equal in distribution.

Let us now consider the parameter space used by the Matrix Normal distribution and the partition of this parameter space induced by the group $\bar{G}$.

Let $\Phi = \{(M, \Sigma_K, \Sigma_D) : M \in K \times D$ matrix, Σ_K is a K x K real, positive definite, symmetric matrix and Σ_D is a D x D real, positive definite, symmetric matrix.$\}$.

The matrix Normal distribution defines equivalency sets on Φ so that $\phi, \tilde{\phi} \in \Phi$ are equivalent if and only if $\tilde{M} = M$ and $\tilde{\Sigma}_D \otimes \tilde{\Sigma}_K = \Sigma_D \otimes \Sigma_K$. Notice that under this model, the parameter combinations (M, Σ_K, Σ_D) and $(M, \frac{1}{c} \Sigma_K, c\Sigma_D)$ for any $c > 0$ are equivalent. The equivalency sets defined under the group $\rightarrow G$ are such that $\phi, \tilde{\phi} \in \Phi$ are equivalent if and only if $\tilde{M} = MR + \underline{1}t^T$ and $\tilde{\Sigma}_D \otimes \tilde{\Sigma}_K = R\Sigma_D R^T \otimes \Sigma_K$. Notice again that here again the parameter combinations (M, Σ_K, Σ_D) and $(M, \frac{1}{c} \Sigma_K, c\Sigma_D)$ for any $c > 0$ are equivalent.

We now consider one particular maximal invariant under this group and derive the parameters that are obtained under it, $T(\cdot)$ and $v(\theta)$ in the notation used before. We then study the partition induced by $v(\theta)$ of the parameter space Θ and discuss the identifiability of various parameters. Since all maximal invariants are equivalent, the identifiability issues will remain the same for all of them.

Let L be a $(K-1) \times K$ matrix whose first column consists of -1's and the rest of the matrix is an identity matrix of dimension $(K-1) \times (K-1)$. Now define $T(X) = LXX^T L^T$. Since $L\underline{1} = \underline{0}$ and because R is an orthogonal matrix i.e., $RR^T = I$, it is easy to see that $T(X) = T(XR + \underline{1}t)$ and therefore is invariant. To show that it is a maximal invariant, we need to show that, given $T(X)$, one can map it back to a unique orbit in the original sample space. This can be proved by using the fact that $T(X)$ is a centered inner product matrix and that there exists a unique mapping (up to rotation, translation and reflection) from the centered inner product matrix to a coordinate matrix, see Lele (1991, 1993). This is also a standard result in the multidimensional scaling literature (Young and Householder, 1938). For an elementary discussion of multidimensional scaling analysis, see Mardia et al., (1979, Chapter 14). Next, we derive the distribution of this maximal invariant in order to determine which parameters are identifiable.

We introduce some additional notation. Let $\Lambda_D = \text{diag}\{\lambda_1, \lambda_2, ..., \lambda_D\}$ denote the diagonal matrix of the eigenvalues of Σ_D and let $\Sigma_K^* = L\Sigma_K L^T$.

Consider a spectral decomposition of the matrix

$$(LM)\Sigma_D^{-1}(LM)^T = \sum_{j=1}^{D} d_j d_j^{~T}$$

where d_j's are the scaled eigenvectors of $(LM)\Sigma_D^{-1}(LM)^T$, that is, $\|d_j\|=\eta_j$, the j-th eigenvalue of $(LM)\Sigma_D^{-1}(LM)^T$. Let $d_j d_j^T = \Delta_j$ and note that Δ_j is of rank one.

Theorem 1: The maximal invariant $T(X)=LXX^TL^T$ is distributed as a linear combination of non-central Wishart matrices. More precisely stated, $T(X) \overset{d}{=} \sum_{j=1}^{D} \lambda_j Z_j$ where $Z_j \overset{d}{=} W_K(1, \Delta_j, \Sigma_K^*)$. a non-central Wishart matrix of dimension $K \times K$ with non-centrality parameter Δ_j and scale parameter Σ_K^*.

Proof: Notice that $LX \sim N(LM, L\Sigma_K L^T, \Sigma_D)$. The above result follows by taking A to be an identity matrix in Theorem 2 presented by deGunst (1987, 248-249).

The partition induced on the parameter space by $T(X)$ is thus given by the inverse image of $(\Lambda_D \otimes L\Sigma_K L^T, \lambda_1 \Delta_1, \lambda_2 \Delta_2, ..., \lambda_D \Delta_D)$. The equivalency sets defined by $v(\theta)$ are such that $\phi, \tilde{\phi} \in \Phi$ are equivalent if and only if $\tilde{M}=MR+\underline{1}t^T$ and $\tilde{\Sigma}_D \otimes \tilde{\Sigma}_K = R\Sigma_D R^T \otimes L\Sigma_K L^T$. Notice that similar to the bivariate case discussed earlier, $\Sigma_K \otimes \Sigma_D$ itself is not identifiable but only $\Lambda_D \otimes L\Sigma_K L^T$ is identifiable. It will be shown in the next section, that the Δ's along with Λ_D can be used to construct the mean form M up to rotation, reflection, and translation. This will be used mainly for representing the mean form graphically. All the inferences would and should be based only on the identifiable parameters. That is, any inference should be the same regardless of which member of the partition induced by $T(X)$ is utilized.

Independence of the partition of the choice of c and L

a) *Choice of c:* Let $v(M, \Sigma_K, \Sigma_D) = (\lambda_1 \Delta_1, \lambda_2 \Delta_2, ..., \lambda_D \Delta_D, \Lambda_D \otimes L\Sigma_K L^T)$. Let $\tilde{\Sigma}_D = c\Sigma_D$ and $\tilde{\Sigma}_K = c^{-1}\Sigma_K$. Then $\tilde{\lambda}_j = c\lambda_j, \tilde{\Delta}_j = c^{-1}\Delta_j$ and $L\tilde{\Sigma}_K L^T = c^{-1}L\Sigma_K LT$ Then $v(M, \tilde{\Sigma}_K, \tilde{\Sigma}_D) = (\tilde{\lambda}_1 \tilde{\Delta}_1, \tilde{\lambda}_2 \tilde{\Delta}_2, ..., \tilde{\lambda}_D \tilde{\Delta}_D, \tilde{\Lambda}_D \otimes L, \tilde{\Sigma}_K L^T) = v(M, \Sigma_K, \Sigma_D)$. The partitions therfore are the same for any $c>0$.

b) *Choice of the centering matrix L:* Let L_1 and L_2 be two different centering matrices. We can write $L_2 = AL_1$ with $A = L_2 L_1^T (L_1 L_1^T)^{-1}$.

To show the invariance of the partition to the choice of the centering matrix, we need to show that for (M, Σ_K, Σ_D) and $(\tilde{M}, \tilde{\Sigma}_K, \tilde{\Sigma}_D)$ in the parameter space Θ, $v_L(M,\Sigma_K,\Sigma_D)=v_L(\tilde{M},\tilde{\Sigma}_K,\tilde{\Sigma}_D)$ if and only if $v_L*(M,\Sigma_K,\Sigma_D)=v_L*(\tilde{M},\tilde{\Sigma}_K,\tilde{\Sigma}_D)$.

Consider first the statement $L_1\Sigma_K L^T_1=L_1\tilde{\Sigma}_K L^T_1$ if and only if $L_2\Sigma_K L^T_2=L_2\tilde{\Sigma}_K L^T_2$.

$$L_1\Sigma_K L^T_1=L_1\tilde{\Sigma}_K L^T_1\Rightarrow AL_1\Sigma_K L^T_1 A^T=AL_1\tilde{\Sigma}_K L^T_1 A^T\Rightarrow L_2\Sigma_K L^T_2=L_2\tilde{\Sigma}_K L^T_2.$$

The converse implication is proved in the same fashion. The rest of the proof follows along the same lines.

Another maximal invariant suggested in the literature is form coordinates (also known as size and shape coordinates; Bookstein, 1986; Kendall, 1989). Its properties are studied in detail by Dryden and Mardia (1991). This maximal invariant does not include reflection as part of the group transformation. Dryden and Mardia (1991) also study the maximal invariant under an additional component of scaling in order to consider only "shape coordinates". The exact distributions of "form" as well as "shape" coordinates are obtained in Dryden and Mardia (1991). Identifiability issues are not dealt with explicitly in any of these papers. More details on the exact distribution of the shape coordinates are discussed later in this chapter. Rao and Suryawanshi (1996, 1998) and Rao (2000) describe a maximal invariant consisting of all possible angles. This is a maximal invariant under the group of transformations which includes scaling.

3.13 Estimation of parameters

Having established the identifiability of certain parameters, the next natural question is whether these parameters can be estimated in a practically suitable and statistically desirable fashion. There have been various methods suggested in the literature. The most commonly used method is based on Generalized Procrustes Analysis. The review article by Goodall (1991) provides a description of this method. Alternatively, maximum likelihood estimation can be conducted using the exact shape distributions derived by Mardia and Dayden (1998) or estimators can be constructed via the method of moments (Stoyan, 1990; Lele, 1993).

Method of moments estimators are based on the moments of the distribution of the maximal invariant described in the previous section. Lele (1993) showed that under the assumption that $\Sigma_D = I$, the estimating functions based on the method of moments have a unique, analytical solution. Moreover, Lele (1993) also shows that the estimator of the mean form M can be obtained up to translation, rotation, and reflection consistently and the covariance parameter $\Sigma_K^* = L\Sigma_K L^T$ is also estimable consistently. However, assuming $\Sigma_D = I$ imposes restrictions on the applicability of this model so we next extend the method to the situation where $\Sigma_D \neq I$.

3.13.1 Method of moments estimators

We will first consider a model where the perturbation of landmarks along the D axes are independent and identical to each other, but the correlations between landmarks are allowed. That is, we consider the perturbation model where the covariance structure is given by $\Sigma_K \otimes I_D$. The main advantage of this model is that there exists a non-iterative, close form, consistent estimator for the mean form matrix and the covariance matrix. For notational simplicity, let $\Sigma_K^* = L\Sigma_K L^T = [\sigma_{lm}]$.

Let e_{lm} denote the squared Euclidean distance between landmarks l and m. Thus $e_{lm} = (X_{l,1} - X_{m,1})^2 + (X_{l,2} - X_{m,2})^2$. Under the Gaussian perturbation model, it can be seen that $(X_{l,1} - X_{m,1})^2$ and $(X_{l,2} - X_{m,2})^2$ are non-central chi-squared random variables with the same scale parameter $\varphi_{lm} = \sigma_{ll} + \sigma_{mm} - 2\varphi_{em}$ and non-centrality parameters $(m_{l,1} - m_{m,1})^2$ and $(m_{l,2} - m_{m,2})^2$. Because the two axes are perturbed independently, it follows that e_{lm} is a sum of two independent non-central chi-squared random variables with common scale parameter. Hence $e_{lm} \sim \varphi_{lm} X_2^2(\varepsilon_{lm}/\varphi_{lm})$ where ε_{lm} is the squared Euclidean distance between landmarks l and m in the true mean form, that is, $\varepsilon_{lm} = (m_{l,1} - m_{m,1})^2 + (m_{l,2} - m_{m,2})^2$ and $\varphi_{lm} = \sigma_{ll} + \sigma_{mm} - 2\varphi_{em}$.

The first two moments of a non-central chi-squared distribution are given by (Johnson and Kotz, 1970, Chapter 28): $E(e_{lm}) = 2\varphi_{lm} + \varepsilon_{lm}$ and

$$\text{var}(e_{lm}) = 4\varphi_{lm}^2 + 4\varepsilon_{lm}\varphi_{lm}. \text{ It thus follows that } \varepsilon_{lm} = \sqrt{E^2(e_{lm}) - \text{var}(e_{lm})}.$$

To obtain estimators we equate the sample moments with the population moments and solve the resulting equations for the parameters.

Let $e_{lm,i}$ denote the squared Euclidean distance between landmarks l and m in the individual "i". Let $\overline{e_{lm}} = \dfrac{1}{n}\sum_{i=1}^{n} e_{lm,i}$ be the average of the squared Euclidean distance between landmarks l and m in n individuals and let $s^2(e_m) = \frac{1}{n}\sum_{i=1}^{n}(e_{lm,i} - \overline{e_{lm}})^2$ be the variance of the squared Euclidean distance between landmarks l and m in n individuals. Then, $\hat{\varepsilon}_{lm} = (\overline{e_{lm}}^2 - s^2(e_{lm}))^{0.5}$ for $l, m = 1, 2, \dots K$.

$$FM(\hat{M}) = \begin{bmatrix} 0 & \sqrt{\hat{\varepsilon}_{12}} & \cdots & \sqrt{\hat{\varepsilon}_{1K}} \\ \sqrt{\hat{\varepsilon}_{12}} & 0 & \cdots & \vdots \\ \vdots & \vdots & \vdots & \vdots \\ \sqrt{\hat{\varepsilon}_{1K}} & \sqrt{\hat{\varepsilon}_{1K-1}} & \cdots & 0 \end{bmatrix}.$$

The estimator of the mean form matrix, which consists of the Euclidean distances between all pairs of landmarks, is thus given by To generalize these results to three-dimensional objects, notice that

$$e_{lm} \sim \varphi_{lm} \; \chi_3^2\left(\frac{\varepsilon_{lm}}{\varphi_{lm}}\right) \quad \text{where} \quad \varepsilon_{lm} = (m_{l,1} - m_{m,1})^2 + (m_{l,2} - m_{m,2})^2 + (m_{l,3} - m_{m,3})^2$$

and $\varphi_{lm} = \sigma_{ll} + \sigma_{mm} - 2\varphi_{em}$. The first two moments for this distribution are given by: $E(e_{lm}) = 3\varphi_{lm} + \varepsilon_{lm}$ and $\mathrm{var}(e_{lm}) = 6\varphi_{lm}^2 + 4\varepsilon_{lm}\varphi_{lm}$. It thus follows that $\varepsilon_{lm} = \sqrt{E^2(e_{lm}) - \frac{3}{2}\mathrm{var}(e_{lm})}$.

In this case, we calculate $\hat{\varepsilon}_{lm} = \sqrt{\overline{e_{lm}}^2 - \frac{3}{2}s^2(e_{lm})}$ for all pairs of

landmarks $l = 1, 2, \dots, K; m = 1, 2, \dots, K$. The estimator of the mean form matrix is obtained by:

$$FM(\hat{M}) = \begin{bmatrix} 0 & \sqrt{\hat{\varepsilon}_{12}} & \cdots & \sqrt{\hat{\varepsilon}_{1K}} \\ \sqrt{\hat{\varepsilon}_{12}} & 0 & \cdots & \vdots \\ \vdots & \vdots & \vdots & \vdots \\ \sqrt{\hat{\varepsilon}_{1K}} & \sqrt{\hat{\varepsilon}_{1K-1}} & \cdots & 0 \end{bmatrix}$$

This matrix may not necessarily correspond to a configuration of K in two- or three-dimensional Euclidean space. This estimator can be improved by constraining it to be a form matrix corresponding to a two- or three-dimensional (as the case may be) object as described below.

STEP 1: Construct the matrix

$$EM(\hat{M}) = \begin{bmatrix} 0 & \hat{\varepsilon}_{12} & \cdots & \hat{\varepsilon}_{1K} \\ \hat{\varepsilon}_{12} & 0 & \cdots & \vdots \\ \vdots & \vdots & \vdots & \vdots \\ \hat{\varepsilon}_{1K} & \hat{\varepsilon}_{1K-1} & \cdots & 0 \end{bmatrix}$$

which is the squared Euclidean distance matrix corresponding to the above mean form matrix.

STEP 2: Calculate the corresponding centered inner product

matrix by $B(\hat{M}) = -\dfrac{1}{2}H[EM(\hat{M})]H^{T}$

where $H = \begin{bmatrix} 1-\dfrac{1}{K} & -\dfrac{1}{K} & \cdots & -\dfrac{1}{K} \\ -\dfrac{1}{K} & 1-\dfrac{1}{K} & \cdots & \vdots \\ \vdots & \vdots & 1-\dfrac{1}{K} & -\dfrac{1}{K} \\ -\dfrac{1}{K} & -\dfrac{1}{K} & \cdots & 1-\dfrac{1}{K} \end{bmatrix}$ is a $K \times K$ symmetric matrix.

STEP 3: Calculate the eigenvalues and eigenvectors of $B(\hat{M})$.
Let us denote its eigenvalues arranged in a decreasing order
by $\lambda_1, \lambda_2, \dots, \lambda_K$ and the corresponding eigenvectors by
$h_1, h_2, \dots, h_K$.

STEP 4: If the original data are from two-dimensional objects,
the estimate of the mean form matrix is given by

$$\hat{M} = [\sqrt{\lambda_1}h_1, \sqrt{\lambda_2}h_2]$$

and if the original data are from three-dimensional objects
the estimate of the mean form matrix is given by

$$\hat{M} = [\sqrt{\lambda_1}h_1, \sqrt{\lambda_2}h_2, \sqrt{\lambda_3}h_3].$$

STEP 5: Calculate the form matrix (the matrix of all pairwise
distances) for the above landmark coordinate matrix. This is
an improved estimator of the mean form matrix.

The matrix M obtained in Step 3 can be used to graphically represent the mean form of the sample. It should be remembered, however,
that this estimator is a representation of the mean form M only up to
rotation, reflection, and translation. In practice, to obtain a graphical
representation that is biologically meaningful, one may need to reflect
the above estimator by multiplying one or more of the axes by -1.

Estimating the covariance matrix $\Sigma_K{}^*$

Having obtained the estimator of the mean form matrix M described above, it is fairly simple to obtain the estimator of the covariance $\Sigma_K{}^*$. Notice that $Y=(LX)(LX)^T\sim Wishart((LM)(LM)^T, L\Sigma_K L^T)$ with D degrees of freedom.

The mean of the non-central Wishart distribution is given by (Arnold, 1981) $E(Y)=(LM)(LM)^T+D(L\Sigma_K L^T)$. Hence, the moment estimator of $\Sigma_K{}^*=L\Sigma_K L^T$ is given by:

$$\hat{\Sigma}_K^* = \frac{1}{D}\left\{\frac{1}{n}\sum_{i=1}^{n} LX_i(LX_i)^T - L\hat{M}(L\hat{M})^T\right\}$$

where $\hat{M}$ is the estimator of the mean form as obtained at the end of Step 4.

The estimator of $\Sigma_K{}^*$ obtained above, although square and symmetric, is not guaranteed to be positive semi-definite. One can obtain a positive semi-definite version using a procedure, sometimes known as Principal Coordinate Analysis. Consider the spectral decomposition of the matrix $\hat{\Sigma}_K{}^*$, namely, $\hat{\Sigma}_K{}^*=PDP^T$ where matrix D is a diagonal matrix with the diagonal elements corresponding to the eigenvalues of $\Sigma_K{}^*$. Replace the negative elements in D by zero and call this modified matrix $\check{D}$. Obtain a new matrix $\hat{\Sigma}_K{}^*=P\check{D}P^T$. This matrix is guaranteed to be square, symmetric, and positive semi-definite.

The estimator given above is slightly different than the one described in Lele (1993). In that paper, instead of $L\Sigma_K L^T$, an estimator for $H\Sigma_K H^T$ was provided. In this monograph, we use the centering matrix L, instead of H, to remain consistent with the rest of the chapter. Replacing L in the above description by H retrieves the formulae and description in Lele (1993).

In the above discussion, we assumed that $\Sigma_D=I$. This imposes some restrictions on the applicability of this model. We now consider a more general situation where the covariance structure is given by $\Sigma_K \otimes \Sigma_D$.

For notational simplicity, let $Y=T(X)=LX(LX)^T$. Let $Y=[Y_{lm}]$ where $l=1,2,\ldots,K; m=1,2,\ldots,K$ denoting the individual elements of the matrix Y. From the previous section, we know that Y is distributed as a linear combination of non-central Wishart random variables with parameters given by $(\Lambda_D \otimes L\Sigma_K L^T, \lambda_1\Delta_1, \lambda_2\Delta_2, \ldots, \lambda_D\Delta_D)$. Let $\Sigma_K{}^*=[\sigma_{lm}]$ and $\Delta_j=[\delta_{lm}{}^j]$. It follows from their definition that the matrices Δ_j's are symmetric and that $\delta_{lm}^j = \sqrt{\delta_{ll}^j \delta_{mm}^j}$.

The following expressions provide the first two moments of the random matrix Y (Alam and Mitra, 1990).

$$E(Y_{lm}) = \sum_{j=1}^{D} \lambda_j \{\delta_{lm}{}^j + \sigma_{lm}\}$$

$$Var(Y_{lm}) = \sum_{j=1}^{D} \lambda_j^2 \{\sigma_{lm}{}^2 + \sigma_{ll}\sigma_{mm} + \sigma_{ll}\delta_{mm}{}^j + \sigma_{mm}\delta_{ll}{}^j + 2\sigma_{lm}\delta_{lm}{}^j\}$$

$$Cov(Y_{lm}, Y_{np}) = \sum_{j=1}^{D} \lambda_j^2 \{\sigma_{ln}\sigma_{mp} + \sigma_{lp}\sigma_{mn} + \sigma_{ln}\delta_{mp}^j + \sigma_{lp}\delta_{mn}^j + \sigma_{mn}\delta_{lp}^j + \sigma_{mp}\delta_{ln}^j\}$$

Without loss of generality, one can assume $\lambda_1 = 1$. Notice that there are $(KD-1) + K(K-1)/2$ unknowns. We choose as many moments and equate the sample moments with the corresponding population moments. Solving the resultant equations, one can numerically obtain the estimates of the parameters $(\lambda_2, \ldots, \lambda_D, \Sigma^*_K, \Delta_1, \Delta_2, \ldots, \Delta_D)$. Given these estimates, we obtain the mean form of the object (up to translation, rotation and reflection). From the definition of Δ_j's, it is clear that they are symmetric matrices with rank 1. One can write $\Delta_j = d_j d_j^T$, where d_j is a scaled eigenvector. If the observations are coming from two-dimensional objects, the estimate of the mean form up to translation, rotation, and reflection is given by $\hat{M} = \begin{bmatrix} 0 & 0 \\ \hat{d}_1 & \sqrt{\hat{\lambda}_2}\hat{d}_2 \end{bmatrix}$,

whereas for three-dimensional objects, the estimate of the mean form up to translation, rotation and reflection is given by

$$\hat{M} = \begin{bmatrix} 0 & 0 & 0 \\ \hat{d}_1 & \sqrt{\hat{\lambda}_2}\hat{d}_2 & \sqrt{\hat{\lambda}_3}\hat{d}_3 \end{bmatrix}.$$

Estimation of the variability around landmarks

In many important biological problems, it is of interest to know the variability around each of the landmarks individually. So far the results described above suggest that it is possible to estimate $L\Sigma_K L^T$ but not Σ_K itself. However, if one is willing to assume a somewhat simpler structure for Σ_K and Σ_D, then it is possible to estimate the variability around each of the landmarks. We assume that the landmarks are perturbed independently of each other, that is, Σ_K as well as Σ_D are diagonal matrices. Without loss of generality, we assume that the first element of Σ_D is equal to 1. We show that for this model, it is possible to obtain the estimator for Σ_K and not just its singular version.

To begin, let us consider the case where $\Sigma_D = I_D$. In the case that Σ_K is diagonal, $L\Sigma_K L^T$ has a typical form that may be exploited to obtain an estimator of Σ_K based on the estimator of $L\Sigma_K L^T$, which we know can be estimated. Let $\Sigma_K = diag(\sigma_1^2, \sigma_2^2, \ldots, \sigma_K^2)$. Then the diagonal elements of $L\Sigma_K L^T$ are given by $\sigma_i^2 + \sigma_1^2$ and all off-diagonal elements are given by $-\sigma_1^2$. Thus a simple estimator of σ_1^2 is given by -1 times the average of all the off-diagonal elements of the estimator of $L\Sigma_K L^T$ obtained in the previous discussion of the general case. The estimators of σ_1^2 are simply obtained by subtracting the average of all the off-diagonal elements of the estimator of σ_1^2 from the i-th diagonal element of the estimator $L\Sigma_K L^T$.

Obviously, these estimators can be improved substantially by using the structure of $L\Sigma_K L^T$ in the method of moments equations directly. The purpose of the above discussion is to show the possibility of estimating the variability around each of the landmarks when certain structures for the covariance matrix are adopted. One can also check the reasonability of the assumed structure by looking at the estimate of $L\Sigma_K L^T$ obtained in the general case and checking if the off-diagonal elements are very similar to each other or not. If they are very similar, the diagonal structure may be a reasonable model.

The consistency of the method of moments estimators follows from the well known fact that sample moments converge to the population moments as the sample size increases (Serfling, 1980). The details can be found in Lele (1993).

3.13.2 Maximum likelihood estimators

Dryden and Mardia (1991) derived the exact distributions of the form coordinates and the shape coordinates. This distribution can be used to write down the likelihood function. One can then maximize this likelihood function to obtain maximum likelihood estimators of the mean form as well as the covariance parameters. It should be noted again that these parameters are identifiable only up to the partition induced by $v(\theta)$. Such a partition was described earlier. A significant drawback of the maximum likelihood approach is that the exact distributions of form and shape coordinates are mathematically complex when the number of landmarks is large and/or the objects are three-dimensional. It is not uncommon in medical research to need at least 10, 15, or more landmarks to represent an object reasonably. In such a situation, the exact shape distribution is extremely complicated as can be seen below. Notice that it involves telescoping sums whose number of components increase geometrically with the number of landmarks.

In the following, we briefly discuss the exact distribution of the shape coordinates, sometimes referred to as the Mardia-Dryden distribution. For the details of the derivation, the reader should refer to Dryden and Mardia (1991,1992). Their paper deals only with two-dimensional objects and we restrict our attention to that case. For generalizations and details, the reader may refer to Dryden and Mardia (1998).

a) Definition of Kendall shape variables

Let X be a $K \times 2$ landmark coordinate matrix and let $X^c = C_I X$ be the centered landmark coordinate matrix where $C_I = diag(H_I, H_I)$ and H_I^T is the usual Helmert matrix without the first column. Notice that this centering matrix is different than the centering matrix L introduced earlier in this chapter. Define new variables as follows:

$$U_i = \frac{(X_{i1}{}^c X_{11}{}^c + X_{i2}{}^c X_{12}{}^c)}{[(X_{11}{}^c)^2 + (X_{12}{}^c)^2]} \text{ and, } V_i = \frac{(X_{i2}{}^c X_{12}{}^c - X_{i1}{}^c X_{12}{}^c)}{[(X_{11}{}^c)^2 + (X_{12}{}^c)^2]}$$

for $i = 1, 2, \ldots, K-2$. Notice that there are $2(K-2)$ variables. These are called Kendall Shape variables. Let us denote the shape variables by a vector S, where $S = \begin{bmatrix} U \\ V \end{bmatrix}$.

b) The probability density function for the Kendall Shape variables

Assume that $vec(X) \sim N(vec(M), \Omega)$.

First define new vectors U^+ and V^+ as follows:

$U^+ = (1, U_1, U_2, \ldots, U_{K-2}, 0, V_1, V_2, \ldots, V_{K-2})$ and $V^+ = (1, -V_1, -V_2, \ldots, -V_{K-2}, 0, U_1, U_2, \ldots, U_{K-2})$. Let $M^* = C\{vec(M)\}, \Omega^* = C\Omega C^T$.

Define new matrices:

$$C = \begin{bmatrix} U^{+T} \Omega^{*-1} U^+ & U^{+T} \Omega^{*-1} V^+ \\ V^{+T} \Omega^{*-1} U^+ & V^{+T} \Omega^{*-1} V^+ \end{bmatrix}, \quad \beta = \begin{bmatrix} M^{*T} \Omega^{*-1} U^+ \\ M^{*T} \Omega^{*-1} V^+ \end{bmatrix}$$

and $g = M^{*T} \Omega^{*-1} U^+ - \beta^T C^{-1} \beta$.

Let $\lambda_1 \geq \lambda_2$ be the eigenvalues of C^{-1} with the corresponding eigenvectors l_1, l_2. Let $\Re_j^{(a)}(x)$ be the generalized Laguerre polynomial of degree j (Abramovitz and Stegun, 1965).

The probability density function of the Kendall shape variables for a two dimensional object is given by:

$$\frac{(K-2)!\lambda_2^{K-2}\exp(-g/2)}{\pi^{K-2}(|C||\Omega^*|)^{1/2}} \sum_{j=0}^{K-2} [\frac{\lambda_1}{\lambda_2}]^j \Re_j^{(-0.5)} \{\frac{-\lambda_1}{2}(\beta^T l_1)^2\} \Re_{K-2-j}^{(-0.5)} \{\frac{-\lambda_2}{2}(\beta^T l_2)^2\} .$$

Notice that the derivation of the distribution does not depend on the Kronecker product form of the variance. Thus this distribution can be used in more general situations than discussed in the methods of moments approach. Although considering the sample sizes encountered in practice, it seems unwise to use a model with such a large number of parameters.

3.13.3 Efficiency comparisons between method of moments and maximum likelihood: a simulations study

All the maximal invariants are equivalent in the sense that maximum likelihood estimators based on any of them give estimates on the same orbit. The use of maximal invariance to eliminate nuisance parameters usually leads to some loss of efficiency as compared to the situation where the nuisance parameters are known. As shown earlier, the method of moments estimators based on the maximal invariant $T(X)$ are easy to obtain. We next study the loss of efficiency of method of moments as compared to maximum likelihood based estimators.

Table 3.5 gives the result of a small simulation study comparing the method of moments based on the inter-landmark differences (EDMA), the MLE based on the Mardia-Dryden distribution and the MLE based on the (unobserved) data before translation or rotation. Thus the "unobservable MLE" represents an idealized situation for comparison since the nuisance parameters, R_i and t_i, are assumed known. Table 3.6 provides the mean form and the covariance structures used for the above simulation study.

The following are the percent relative root mean squared errors for the two methods based on 100 simulations. Samples of size 30 were generated under two different mean forms and two different covariance structures (see Table 3.6). The column "Unobserved MLE" corresponds to the relative root mean squared error for the maximum likelihood estimators based on the assumption the nuisance parameters are known. These values represent best achievable results.

Table 3.5 Efficiency Comparisons Between the Method of Moments and the Method of Maximum Likelihood.

Scenarios	Parameters	EDMA	MLE	Unobserved MLE
$(M_1, \Sigma_{K,1}, I_2)$	Mean distances*	1.0	1.0	1.0
	Variances*	23.2	19.5	16.3
	Covariances*	45.6	37.7	28.4
$(M_1, \Sigma_{K,2}, I_2)$	Mean distances	1.5	1.5	1.5
	Variances	21.4	18.7	14.6
	Covariances	36.4	31.0	24.7
$(M_2, \Sigma_{K,1}, I_2)$	Mean distances	2.2	2.2	2.2
	Variances	24.2	19.4	14.9
	Covariances	48.4	41.3	30.1

* Because of the non-convergence of the maximum likelihood routine, these are based on only 98 out of 100 simulations.

Table 3.6 The Mean Form Coordinates (centered) and the Covariance Matrices Used in the Simulation Study.

$$\Sigma_{K,1} = \begin{bmatrix} 0.78 & 0.44 & 0.28 \\ 0.44 & 0.66 & 0.40 \\ 0.28 & 0.40 & 0.53 \end{bmatrix} \quad \Sigma_{K,2} = \begin{bmatrix} 0.78 & 0 & 0 \\ 0 & 0.66 & 0 \\ 0 & 0 & 0.53 \end{bmatrix}$$

$$M_1 = \begin{pmatrix} 0 & 4.72 \\ 7.07 & -2.36 \\ -7.07 & 2.36 \end{pmatrix} \quad M_2 = \begin{pmatrix} -1 & 2.33 \\ 4 & -0.67 \\ -3 & -1.66 \end{pmatrix}$$

In estimating the inter-landmark distances or the variance-covariance parameters, none of the methods exhibited any bias, therefore Table 3.5 concentrates on the relative root mean square error, which primarily reflects variability. Since performance of the methods was similar within the mean parameters, within the variance parameters and within the covariance parameters, the results are averaged across the (three in each case) parameters. We can see that, for estimating the mean form, both EDMA and the MLE do as well as the unobservable MLE. For estimating the variance parameters EDMA and the MLE perform worse than the unobservable MLE, and EDMA performs slightly worse than the MLE. For the covariance parameters, there is some loss of efficiency in using EDMA compared to the MLE, but not a dramatic amount.

Another point is worth mentioning in comparing EDMA with the MLE: even for the simple case of three landmarks in two-dimensions, we found the likelihood very difficult to maximize numerically using the Dryden and Mardia (1991) distribution. Despite estimating the variance-covariance matrix using a Cholesky decomposition to guarantee positive semi-definiteness, many data sets required new starting values and/or a preliminary sizing step for the variance-covariance parameters in order to achieve convergence. Despite repeated attempts there were still 2 data sets (out of 100) in the first parameter configuration for which we could not obtain convergence.

The practical utility of the maximum likelihood method of estimation based on Mardia-Dryden distribution is therefore somewhat suspect, especially when one is dealing with a large number of landmarks for three-dimensional objects and a medium sample size. Such a situation is not at all uncommon and should be clear from the typical data analyses presented in this book as well as various papers cited herein.

As a practical alternative to the maximum likelihood estimation, Goodall and Bose (1987) and Goodall (1991) discuss Procrustes superimposition based estimators of the mean form, mean shape, and the variance-covariance matrix. See also Bookstein (1991) for similar estimators based on Bookstein shape coordinates. Kent (1994) and Kent and Mardia (1997) show that the Procrustes estimators can also be viewed as estimators based on tangent space approximation to the Kendall shape space. These estimators are computationally easier than the maximum likelihood estimators. Unfortunately, these estimators, under the Gaussian perturbation model with general covariance structure, are inconsistent for mean form as well as mean shape (Lele, 1993, Kent and Mardia, 1997).

Moreover, simulations reported in Lele (1993) and the discussion in Part 1 of this chapter show that the Procrustes estimator of the variability does not correspond to the variance-covariance matrix used in the Gaussian perturbation model. Given these results, we find the Procrustes estimators inadequate. They are not discussed here in detail. See Dryden and Mardia (1998) for a detailed discussion of the Procrustes and related approaches.

3.14 Computational algorithms

3.14.1 Algorithms for estimation of mean and variance parameters

In the following, we provide detailed computational algorithms for estimation of the mean form matrix and variance-covariance matrix. In order to program these algorithms, the reader will need to understand the following concepts and related computation procedures: Form matrix, Centered landmark coordinate matrix, Centered inner product matrix, Eigenvalues, and Eigenvectors. Except for the concepts of eigenvalues and eigenvectors, all other concepts are described below. For details on eigenvalues and eigenvectors of a square, symmetric matrix, refer to any standard matrix algebra textbook such as Barnett (1990). For computational algorithms, see Press et al., (1986).

Definition of a centered landmark coordinate matrix: Let A be a landmark coordinate matrix. First, calculate the mean of the first column and subtract it from all the elements of the first column; calculate the mean of the second column and subtract it from all the elements of the second column, and calculate the mean of the third column and subtract it from all the elements of the third column. The resultant matrix, denoted by A^c, is called the centered landmark coordinate matrix.

For example, let the landmark coordinate matrix be $A = \begin{bmatrix} 0 & 0 \\ 1 & 0 \\ 0 & 1 \end{bmatrix}$.

Then the centered landmark coordinate matrix is given by subtracting from each of the entry in a column by the average of the corresponding column in the landmark coordinate matrix A. Thus

$$A^c = \begin{bmatrix} -0.33 & -0.33 \\ 0.66 & -0.33 \\ -0.33 & 0.66 \end{bmatrix}$$ is the centered landmark coordinate matrix.

Notice that the column sum of A^c is always 0. Notice that this matrix is obtained when one shifts the original triangle such that its centroid, instead of matching landmark 1, matches with (0,0).

Centered inner product matrix: Let A^c be a centered landmark coordinate matrix. Define a new matrix $B=A^c(A^c)^T$. This matrix is called the centered inner product matrix corresponding to A.

The centered inner product matrix corresponding to the above example is given by

$$B(A) = A^c(A^c)^T = \begin{bmatrix} 0.218 & 0.112 & -0.112 \\ 0.112 & 0.558 & -0.442 \\ -0.112 & -0.442 & 0.558 \end{bmatrix}$$

Algorithm 1

This algorithm describes the estimation procedure when the covariance structure is $\Sigma_K \otimes I_D$, similar to the covariance structure depicted in Figure 3.1d. The estimation procedure is particularly simple in this situation. We suggest that the reader first try to program this algorithm and then try the more general algorithm (Algorithm 2) that is applicable when the covariance structure is of the form $\Sigma_K \otimes \Sigma_D$. All the data analyses carried out in this book and various papers using Euclidean Distance Matrix Analysis (EDMA) use Algorithm 1 as presented by Lele (1993). The parameters estimated are the form matrix, $FM(M)$, corresponding to the mean form M and the covariance matrix $\Sigma_K{}^*$, the singular version of Σ_K.

Let $X_1, X_2, \ldots, X_n$ denote the landmark coordinate matrices of n individuals in the sample.

a) Algorithm for estimation of mean form:

STEP 1: Calculate the form matrix corresponding to each of the landmark coordinate matrices $X_1, X_2, \ldots, X_n$. For simplicity of notation let us denote the form matrices by F_i instead of the clumsier (albeit clearer) notation FM_i or $FM(X_i)$. Thus F_i is a matrix of Euclidean distances between all pairs of landmarks for the i-th individual.

STEP 2: Calculate the squared Euclidean distance matrices $E_1, E_2, \ldots, E_n$ corresponding to each individual in the sample, where E_i is obtained by squaring each element in the matrix F_i. Let $e_{lm,i}$ denote the squared Euclidean distance between landmarks l and m in the individual "i".

STEP 3: Calculate $\overline{e_{lm}} = \dfrac{1}{n}\sum_{i=1}^{n} e_{lm,i}$, the average of the squared

Euclidean distance between landmarks l and m in n individuals.

STEP 4: Calculate $s^2(e_{lm}) = \dfrac{1}{n}\sum_{i=1}^{n}(e_{lm,i} - \overline{e_{lm}})^2$, the variance of the squared Euclidean distance between landmarks l and m in n individuals.

STEP 5: If the original objects are two-dimensional, calculate $\hat{\varepsilon}_{lm} = (\overline{e_{lm}}^2 - s^2(e_{lm}))^{0.5}$ for all pairs of landmarks. If the original objects are three dimensional, calculate

$\hat{\varepsilon}_{lm} = (\overline{e_{lm}}^2 - \dfrac{3}{2}s^2(e_{lm}))^{0.25}$ for all pairs of landmarks

$l = 1,2,\ldots,K; m = 1,2,\ldots,K.$

The estimate of the mean form matrix, that is the estimate of the distances between pairs of landmarks in the mean form (the red pen transparency) is provided by

$$FM(\hat{M}) = \begin{bmatrix} 0 & \sqrt{\hat{\varepsilon}_{12}} & \cdots & \sqrt{\hat{\varepsilon}_{1K}} \\ \sqrt{\hat{\varepsilon}_{12}} & 0 & \cdots & \vdots \\ \vdots & \vdots & \vdots & \vdots \\ \sqrt{\hat{\varepsilon}_{1K}} & \sqrt{\hat{\varepsilon}_{1K-1}} & \cdots & 0 \end{bmatrix}$$

This estimator can be improved by constraining it to be a form matrix corresponding to a two- or three-dimensional (as the case may be) object as described below.

STEP 6: Construct the matrix

$$EM(\hat{M}) = \begin{bmatrix} 0 & \hat{\varepsilon}_{12} & \cdots & \hat{\varepsilon}_{1K} \\ \hat{\varepsilon}_{12} & 0 & \cdots & \vdots \\ \vdots & \vdots & \vdots & \vdots \\ \hat{\varepsilon}_{1K} & \hat{\varepsilon}_{1K-1} & \cdots & 0 \end{bmatrix}$$

the squared Euclidean distance matrix corresponding to the above mean form matrix.

STEP 7: Calculate the corresponding centered inner product matrix by $B(\hat{M}) = -\dfrac{1}{2}H[EM(\hat{M})]H^T$ where

$$H = \begin{bmatrix} 1-\dfrac{1}{K} & -\dfrac{1}{K} & \cdots & -\dfrac{1}{K} \\ -\dfrac{1}{K} & 1-\dfrac{1}{K} & \cdots & \vdots \\ \vdots & \vdots & 1-\dfrac{1}{K} & -\dfrac{1}{K} \\ -\dfrac{1}{K} & -\dfrac{1}{K} & \cdots & 1-\dfrac{1}{K} \end{bmatrix}$$ is a $K \times K$ symmetric matrix.

STEP 8: Calculate the eigenvalues and eigenvectors of $B(\hat{M})$. Let us denote its eigenvalues arranged in a decreasing order by $\lambda_1, \lambda_2, ..., \lambda_K$ and the corresponding eigenvectors by $h_1, h_2, ..., h_n$.

STEP 9: If the original data are from two-dimensional objects, the estimate of the mean form matrix is given by

$\hat{M} = [\sqrt{\lambda_1}h_1, \sqrt{\lambda_2}h_2]$, and if the original data are from three-dimensional objects, the estimate of the mean form matrix is given by $\hat{M} = [\sqrt{\lambda_1}h_1, \sqrt{\lambda_2}h_2, \sqrt{\lambda_3}h_3]$.

STEP 10: Calculate the form matrix (the matrix of all pairwise distances) for the above landmark coordinate matrix. This is an improved estimator of the mean form matrix.

Note: The M obtained in STEP 3 can be used to graphically represent the mean form of the sample. It should be remembered that this estimator is a representation of the mean form M only up to rotation, reflection, and translation. In practice, to obtain a graphical representation that is biologically meaningful, one may need to reflect the above estimator by multiplying one or more of the axes by -1.

Having obtained the estimator of the mean form matrix M described above, it is fairly simple to obtain the estimator of the covariance matrix, Σ_K^*. The following steps provide the details.

b) Algorithm for estimation of Σ_K^*

STEP 1: Calculate the centered matrices $X_1^c, X_2^c, ..., X_n^c$ corresponding to the observations $X_1, X_2, ..., X_n$.

STEP 2: Calculate $\hat{\Sigma}_K^* = \frac{1}{D}\{\frac{1}{n}\sum_{i=1}^{n} X_i^c (X_i^c)^T - \hat{M}\hat{M}^T\}$ where D is the dimension of the observations.

The estimator obtained above, although square and symmetric, is not guaranteed to be positive semi-definite. One can obtain a positive semi-definite version using the following steps.

STEP 3: Calculate the spectral decomposition of the matrix $\hat{\Sigma}_K^*$. That is, find its eigenvectors and eigenvalues and write it as $\hat{\Sigma}_K^* = PDP^T$ where matrix D is a diagonal matrix with the diagonal elements corresponding to the eigenvalues of Σ_K^*. If there are any negative elements in D, replace them by zero and call this modified matrix $\tilde{D}$. Obtain a new matrix $\Sigma_K^* = P\tilde{D}P^T$. This matrix is guaranteed to be square, symmetric, and positive semi-definite.

Algorithm 2

In this section, we provide an algorithm for the estimation of the mean form and the covariance structure when we expect that the perturbations along the x, y, and z axes are correlated with each other. We provide the algorithm for the two-dimensional three landmarks case in detail. More general situations can be developed using the description in Part 2. The reader should, however, be aware that to fit more complex models, larger sample sizes are needed to obtain reasonable estimates. Thus, although more realistic, in practice, this general model might be more difficult to fit.

Define a matrix $\quad L = \begin{bmatrix} -1 & 1 & 0 \\ -1 & 0 & 1 \end{bmatrix}$.

Let $X^c = LX$ denote the centered matrices. This centering is slightly different than the one used in the previous algorithm.

Let $\Sigma_K = L\Sigma_K L^T$ and Λ_D denote a diagonal matrix consisting of the eigenvalues of Σ_D.

It was proven in Part 2 that these are the only estimable quantities related to covariance.

Define a new matrix $\quad \Delta_j = \begin{bmatrix} \delta_{11}{}^j & \sqrt{\delta_{11}{}^j \delta_{22}{}^j} \\ \sqrt{\delta_{11}{}^j \delta_{22}{}^j} & \delta_{22}{}^j \end{bmatrix}$ for $j = 1,2$.

STEP 1: Calculate the centered matrices $X_1{}^c, X_2{}^c, \ldots, X_n{}^c$ where $X_i{}^c = L X_i$.

STEP 2: Calculate the centered inner product matrices $B_i = X_i{}^c (X_i{}^c)^T$. Notice that these centered inner product matrices are 2×2 symmetric matrices. Let $B_{11,i}, B_{22,i}, B_{12,i}$ be the three distinct elements of B_i.

STEP 3: Calculate the following quantities:

$$B_{11} = \frac{1}{n} \sum_{i=1}^{n} B_{11,i}$$

$$B_{22} = \frac{1}{n} \sum_{i=1}^{n} B_{22,i}$$

$$B_{12} = \frac{1}{n} \sum_{i=1}^{n} B_{12,i}$$

$$\mathrm{var}(B_{11}) = \frac{1}{n} \sum_{i=1}^{n} (B_{11,i} - B_{11})^2$$

$$\mathrm{var}(B_{22}) = \frac{1}{n} \sum_{i=1}^{n} (B_{22,i} - B_{22})^2$$

$$\mathrm{var}(B_{12}) = \frac{1}{n} \sum_{i=1}^{n} (B_{12,i} - B_{12})^2$$

$$\mathrm{cov}(B_{11}, B_{22}) = \frac{1}{n} \sum_{i=1}^{n} (B_{11,i} - B_{11})(B_{22,i} - B_{22})$$

$$\mathrm{cov}(B_{11,} B_{12}) = \frac{1}{n} \sum_{i=1}^{n} (B_{11,i} - B_{11})(B_{12,i} - B_{12})$$

$$\mathrm{cov}(B_{22}, B_{12}) = \frac{1}{n} \sum_{i=1}^{n} (B_{22,i} - B_{22})(B_{12,i} - B_{12})$$

STEP 4: Solve the following sets of equations simultaneously under the constraints: $\lambda=1$ and $\delta_{12}{}^j = \sqrt{\delta_{11}{}^j \delta_{22}{}^j}$ for $j = 1,2$. Notice that there are eight equations and eight unknowns.

$$\sum_{j=1}^{2} \lambda_j \{\delta_{11}^j + \sigma_{11}\} = B_{11}$$

$$\sum_{j=1}^{2} \lambda_j \{\delta_{22}^j + \sigma_{22}\} = B_{22}$$

$$\sum_{j=1}^{2} \lambda_j \{\delta_{12}^j + \sigma_{12}\} = B_{12}$$

$$\sum_{j=1}^{2} \lambda_j^2 \{\sigma_{11}^2 + \sigma_{11}\sigma_{11} + \sigma_{11}\sigma_{11}^j + \sigma_{11}\sigma_{11}^j + 2\sigma_{11}\sigma_{11}^j\} = \text{var}(B_{11})$$

$$\sum_{j=1}^{2} \lambda_j^2 \{\sigma_{22}^2 + \sigma_{22}\sigma_{22} + \sigma_{22}\delta_{22}^j + \sigma_{22}\delta_{22}^j + 2\sigma_{22}\delta_{22}^j\} = \text{var}(B_{22})$$

$$\sum_{j=1}^{2} \lambda_j^2 \{\sigma_{12}^2 + \sigma_{11}\sigma_{22} + \sigma_{11}\delta_{11}^j + \sigma_{22}\delta_{22}^j + 2\sigma_{12}\delta_{12}^j\} = \text{var}(B_{12})$$

$$\sum_{j=1}^{2} \lambda_j^2 \{\sigma_{12}\sigma_{12} + \sigma_{12}\sigma_{12} + \sigma_{12}\delta_{12}^j + \sigma_{12}\delta_{12}^j + \sigma_{12}\delta_{12}^j + \sigma_{12}\delta_{12}^j\} = \text{cov}(B_{11}, B_{22})$$

$$\sum_{j=1}^{2} \lambda_j^2 \{\sigma_{11}\sigma_{12} + \sigma_{12}\sigma_{11} + \sigma_{11}\delta_{12}^j + \sigma_{12}\delta_{11}^j + \sigma_{11}\delta_{12}^j + \sigma_{12}\delta_{11}^j\} = \text{cov}(B_{11}, B_{12})$$

STEP 5: The estimate of the coordinates of the mean form is given by:

$$\hat{M} = \begin{bmatrix} 0 & 0 \\ \delta_{11}^{\;1} & \delta_{11}^{\;2} \\ \sqrt{\lambda_2}\delta_{22}^{\;1} & \sqrt{\lambda_2}\delta_{22}^{\;2} \end{bmatrix}.$$

This can be used to calculate the form matrix by calculating the distances between the three landmarks.

The estimate of the covariance matrix $\Sigma_K^* = L\Sigma_{KL}L^T$ is given by

$\hat{\Sigma}_K^* = \begin{bmatrix} \sigma_{11} & \sigma_{12} \\ \sigma_{12} & \sigma_{22} \end{bmatrix}$ and the estimate of the eigenvalues of

Σ_D is given by $\Lambda_D = \begin{bmatrix} 1 & 0 \\ 0 & \lambda_2 \end{bmatrix}.$

Generalization of this algorithm to three-dimensional objects and objects with more than 3 landmarks is notationally complicated but straightforward. We do not provide the details here.

3.14.2 Generating observations using the estimated mean form and the covariance matrix

We will describe the algorithm for the covariance model $\Sigma_K \otimes I_D$ in detail. This is the model that is extensively used throughout this monograph. The following algorithm is described for a two-dimensional object. Generalization to three-dimensional object is straightforward.

Let M_{K-1} be a submatrix consisting of the first $(K-1)$ rows of M, the mean form coordinate matrix obtained in Step of Algorithm 1. Let Σ_{K-1}^* denote the $(K-1)\times(K-1)$ submatrix of Σ_K^*. This matrix consists of the first $(K-1)$ rows and $(K-1)$ columns of Σ_K^*. In general, Σ_{K-1}^* should be a positive definite matrix. The non-positive definiteness of Σ_{K-1}^* implies that the sample size is much too small to fruitfully conduct any statistical analysis, see the discussion in Part 1 of this chapter.

Algorithm for generating matrix normal random variates

STEP 1: Obtain the Cholesky decomposition of Σ_{K-1}^*. That is, obtain an upper triangular matrix C such that $\Sigma_{K-1}^* = CC^T$.
STEP 2: Generate $2(K-1)$ random numbers from $N(0,1)$ distri-

bution and arrange them in a $(K-1)\times2$ matrix. Call this matrix Z.

STEP 3: Calculate $Y = M_{K-1} + CZ$. The matrix Y is a $(K-1)\times2$ matrix.

STEP 4: Construct a $K\times2$ matrix X by adding the K-th row to the Y matrix such that the sum of each column is zero. That is, the K-th row element of a column is obtained by simply adding the preceding $(K-1)$ elements and multiplying the sum by -1. The matrix X so obtained is a realization from the matrix normal distribution with mean M and covariance $\Sigma_K^* \otimes I_D$.

STEP 4: Repeat STEPS 1-3 as many number of times as the required number of observations.

For a three dimensional object, replace the STEP 2 above by:

STEP 2: Generate $3(K-1)$ random numbers from $N(0,1)$ distribution and arrange them in a $(K-1)\times3$ matrix.

CHAPTER 4

Statistical Methods for Comparison of Forms

"In a very large part of morphology, our essential task lies in the comparison of related forms rather than in the precise definition of each...."

D'Arcy Thompson (1992)

The previous chapter presented our notation and method for representing the form of an object and mean form for a sample. In this chapter, we turn to the comparison of forms and introduce our method, Euclidean Distance Matrix Analysis, or EDMA (pronounced ed·ma). To clarify our approach, we consider other methods for the comparison of forms, all of which can be classified as either superimposition techniques or deformation techniques. We then present the EDMA approach to the comparison of forms, providing suggestions for interpreting the output of these analyses and statistical approaches. Statistical testing of the equality of shapes is provided in keeping with traditional aspects of statistics. Since the testing of similarity in shapes provides little in the way of biological information, we de-emphasize that aspect of statistical testing and stress the importance of exploring and demonstrating local differences between forms using EDMA output (i.e., the form difference matrices) and confidence interval methods. Finally, we provide analyses of our example data sets in order to clarify and solidify the ideas presented.

4.1 Introduction

When observing the form of any object, we may be struck with a curiosity about why the object looks the way it does and why it is different from the things around it. Consciously or unconsciously, we compare one object to others to try to explain or justify its appearance. When observing the form of any biological organism, two key questions come to mind. First, why do organisms look the way they do? And second, why do organisms look different from one another? Members of biological groups are often distinguished from one another on the basis of appearance or morphology. Morphologists have always suspected that form follows function to some degree (Cuvier, 1828; Hildebrand et al., 1985; Radinsky, 1987) but have sought explanations for form using not only biomechanical reasoning and functional models but also phylogenetic information, optimization criteria, consideration of sexual dimorphism, developmental programs, genetic factors, evolutionary trends, and mathematical models. To understand the relationship between form and any of these explanatory variables, a method is needed that enables both a precise definition of form and the quantitative comparison of forms, including the ability to "localize" form differences.

In practice, there are several limiting factors that make something as seemingly simple as comparing forms extraordinarily difficult. In the following section, we discuss the implications of these limiting factors for some of the more popular methods of superimposition and deformation by examining several Procrustean approaches (e.g., Rohlf and Slice, 1990; Goodall, 1991; Dryden and Mardia, 1998) and the use of thin plate splines (Bookstein, 1989; Bookstein, 1991). The problem of invariance discussed previously in the context of describing form is also encountered when comparing forms. We show that the shortcomings of superimposition and deformation approaches stem from their failure to satisfy the principle of invariance. EDMA is shown to satisfy the principle of invariance. Sections 4.2 through 4.6 are based on ideas originally presented in Lele (1999).

4.2 Limiting factors in morphometrics

The first limiting factor of any morphometric analysis concerns the type of data chosen for analysis. With the choice of landmark coordinate data comes the acknowledgment that landmarks can never entirely represent a biological form or the form of the population that the specimen represents. This issue was more fully discussed in Chapter 2. The decision to use landmark data to represent a form is a primary limitation that is accepted and understood. Additional limiting factors faced by morphometricians are divided in to three major classes introduced below. These include restricted information, biological variability, and lack of a common coordinate system.

1. **Incomplete information:** Landmark data used to study form can be restricted in two ways.

   (a) Sample size: In practice, it may be difficult to get a large number of observations pertaining to the problem under study. For example, the number and condition of available fossil specimens may be limited, or the disease being investigated may be too rare to obtain many cases. In certain cases, the "normal" sample used for comparison can also be limited in size.

   (b) Number and choice of landmarks: The form of biological objects is extremely complex. Our approach to the study of form involves choosing biologically relevant loci to serve as landmarks. There may be situations in which one region or portion of a form contains more closely spaced landmarks than another portion of the same form, or where certain regions are without landmarks. This results in unequal representation of aspects of a single form. One important criterion for choosing landmarks is that they reflect positions representative of underlying processes. This and issues of repeatability guide our decision whether to include any specific landmark in the analysis. Consideration of both of these concerns places limits on the number and nature of landmarks available for analysis.

2. **Variability:** Variability is ubiquitous in nature and is of great interest to biologists, but it is also our nemesis. In the previous chapter we discussed the presence of variability among biological organisms, the need to correctly account for variability, and ways to model within-population variability. We noted that the nuisance parameters of translation, rotation, and reflection contribute to the difficulty of correctly estimating variability. These same nuisance parameters contribute to the difficulties in comparing samples of forms. These issues are explored more fully in this chapter.

3. **Lack of a common coordinate system:** The presence of nuisance parameters is linked to the lack of a common coordinate system. The effect of these nuisance parameters on the estimation of mean form and variability were documented in the previous chapter. Though often unappreciated by basic

research scientists and statisticians, the lack of a common coordinate system has severe implications when comparing forms. This is because methods that are coordinate system based require that a particular coordinate system be adopted, and this choice can profoundly affect analytical results. It is important that these issues and their implications be examined and understood thoroughly.

4.3 Comparing two forms: introduction to the problem

When comparing samples of biological forms, within-population variability cannot be ignored. However, for instructional purposes we begin our discussion by comparing two forms rather than two samples of forms. We do this to clarify the act of form comparison and to isolate and observe the effect of the lack of a common coordinate system on the comparison of forms. The consequence of within-population variability for the comparison of forms is discussed later in this chapter.

Suppose we are studying two groups of organisms (e.g., two different species, sexes, or age groups) by choosing a representative form from each of the groups. Typically, the first question posed is: are the two forms different? It is important to have quantitative data that substantiate the answer to this question, especially when form differences are subtle. However, many times the differences are substantial and the answer is obviously affirmative. The more important question then becomes: *in what aspects* do the two forms differ from one another? The answer to this question requires identifying those loci where differences are profound and those where differences are minimal. We refer to this process as the "localization of form differences." Information pertaining to the location of component parts that are most or least different between specimens is essential to the investigator's discovery of the mechanisms (e.g., physiological, genetic, developmental, pathological, environmental, evolutionary) that underlie the form difference or the form change.

Before delving into the comparison of complex biological forms, let us first consider a simple triangle in two dimensions with three landmarks. We will conduct a simple experiment using three circular transparencies with different diameters as the media on which data are recorded. Start by drawing the three points of a triangle on one of the transparencies using a red pen. Number the points from 1 to 3 and think of this as the average triangle for the first group. Put another transparency on top of the first and, using a green pen, place three points on the second transparency in close proximity to the red points on the first transparency, but change the location of each of the points any way you like. Label these points from 1 to 3 corresponding to the numbering system for the first triangle. Think of the triangle that you just drew in green as the average of a second group.

Now, take a third transparency and lay it over the first two without disturbing the relative positions of the points drawn in red and in green on the two original transparencies. On the third transparency, draw vectors connecting the landmark drawn in red and labeled "1" to the landmark drawn in green and labeled "1," then connect the red landmark "2" to the green landmark "2," and finally connect the red landmark "3" to the green landmark "3," with a vector. These vectors, originating at the red point and ending at the green point, justifiably represent the form difference or form change required to go from triangle 1 to triangle 2, providing precise information pertaining to the direction and magnitude of change local to each landmark. The transparency that contains these vectors represents the "true form change." The vectors enable "localization" of the changes from one form to the other.

We refer to the transparency with red markings as the "red transparency," the transparency with the green markings as the "green transparency," and the transparency with the vectors depicting the change from red to green as the "true form change" transparency. Take away the true form change transparency and set it aside for future use. Move the red and green transparencies away from one another in an arbitrary fashion (e.g., throw them on the floor and pick them up). This exercise simulates the effect of the loss of a common coordinate system. Once the common coordinate system is lost, our job is to determine whether the vectors that depict the true form change can be recreated using only the landmark data recorded on the red and green transparencies.

The conditions we have described in this experiment reflect the situation of the scientist who wants to compare two objects. All that the scientist has are the red and green transparencies. Nothing inherent to those transparencies provides information regarding the common coordinate system that enabled the recording of the true form change. Superimposition approaches posit that the true form change depicted as vectors that represent the magnitude and direction of change can be found using these red and green transparencies. Next, we present superimposition approaches in detail to gain an understanding of why this claim is invalid.

4.4 Superimposition-based approaches

The most intuitive way to compare two objects is to overlay one on top of the other, or to superimpose them, in order to visualize the differences between them. Many scientific approaches to the study of form have developed from this simple idea. Boas (Boas, 1905) proposed a superimposition approach at the beginning of the twentieth century, although his role in the establishment of this approach for the field of anthropology was not broadly recognized until recently (Cole, 1996). The invention of x-ray techniques and the consequent development of Roentgen cephalometry (Broadbent, 1975) early in this century spawned a number of research projects focusing on the collection and analysis of longitudinal data. Radiology permitted the study of growth of internal structures and the field of Roentgen cephalometry grew out of the simple comparison of head x-rays by overlaying two-dimensional tracings of the pertinent data. The first comparisons were done by manually matching x-rays to the center of the pituitary fossa (marked by a landmark called "sella") and orienting on a line that stretched from sella to a point marking the intersection of the nasal bones with the frontal bone (the landmark nasion; see Figure 4.1). Methods of standardization and various quantitative approaches were later developed so that comparisons could be made between data collected from various longitudinal series using more automated techniques (Nanda, 1956). Additionally, two- and three-dimensional superimposition-based approaches have been proposed to determine localized differences between forms. For example, a recent comparison of cranial profiles of evolving lineages of *Homo* used Procrustes superimposition of landmarks and semi-landmarks located on the internal surface of the frontal bone (Bookstein et al., 2000).

The fundamental idea behind superimposition-based approaches is quite intuitive. The following steps relate closely to those presented in the transparency experiment and explain the basic idea behind all superimposition approaches. Fix the first object (the red transparency). Translate and rotate the second object (the green transparency) so that the landmarks on the green transparency lie as close as possible to the corresponding landmarks on the red transparency. Then draw the vectors that connect the red landmark 1 to the green landmark 1,

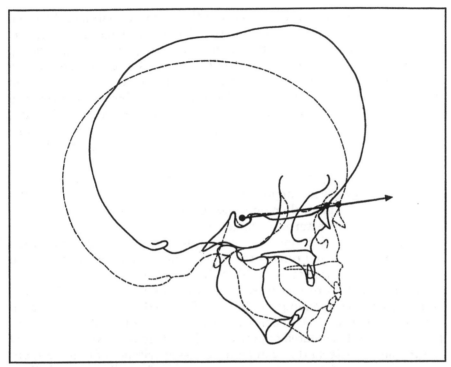

Figure 4.1. Comparison of the tracing of a cephalometric radiograph of a single individual with Apert syndrome (age 2 ¹/₂ years, solid line) and an unaffected individual (same approximate age, broken line). The two tracings are superimposed on the landmark sella and oriented to a line that connects the landmarks sella and nasion. From this superimposition, interpretations are made regarding how the skulls of unaffected and affected individuals differ from one another.

the red landmark 2 to the green landmark 2, and so on. The vectors drawn are presumed to depict the true form change and are used to infer biologically based reasons for the changes/differences observed.

In the instructions given above, we asked that you match all corresponding landmarks to the best of your ability. Methodologically, various algorithms are used in superimposition schemes. Each algorithm is designed to minimize a particular measure of similarity (e.g., Sneath, 1967; Gower, 1975; Bookstein, 1978; Siegel and Benson, 1982; Bookstein, 1986; Goodall and Bose, 1987). We refer to the way in which forms are superimposed so as to minimize a particular measure as the "minimization criterion." A commonly used measure of similarity is based on the Procrustes distance where the minimization criterion includes fixing one object and superimposing other objects on that one in an attempt to minimize the sum of the squared distances between corresponding landmarks on the forms. In our transparency example, this process corresponds to fixing the red transparency and rotating and translating the green transparency so that the sum of the squared distances between corresponding landmarks on the red and green transparencies is minimized (see Goodall, 1991; Dryden and Mardia, 1998 for details on the "Generalized Procrustes Analysis" (GPA) approach to superimposition). The application of GPA to the comparison of our triangles results in the difference between the red and green transparencies being roughly equally distributed among all landmarks.

Methodologically, the measure of similarity (the minimization criterion) differs between approaches. This follows intuitively from the fact that there is more than one way to superimpose. In practice, minimization of an alternate measure will produce an alternate superimposition of forms. For example, another superimposition scheme translates and rotates the green transparency so that the sum of the absolute distances (instead of the sum of the squared distances) is minimized. This results in some landmarks being matched closely, while other corresponding landmarks fall farther from one another. Siegel and Benson (1982) describe this "Robust Procrustes" superimposition and other related approaches. These are only two examples of minimization criteria that can be used to superimpose forms. In reality, there are infinite potential measures of similarity and their corresponding superimposition schemes, although some may be more sensible than others. Figure 4.2 illustrates the results of superimposing our example triangles according to various schemes. Figure 4.2a shows the "true form difference" that we created (the form difference that you created in the experiment will be different), Figure 4.2c shows the form difference depicted by the robust superimposition method, and Figure 4.2b shows the form difference depicted by the least squares superimposition method. Figure 4.2d provides the form difference depicted by edge superimposition. Notice that none of the schemes recreate the "true form difference," and that the various superimposition schemes provide varying versions of the overall comparison with dissimilar descriptions of the local differences between the red and green transparency.

Once the original relative positions of the red and green transparencies are disturbed, we lose the information necessary to put them back into their original, relative positions. The failure of superimposition schemes in localizing the true form change stems from the fact that once disturbed, the original relationship between the transparen-

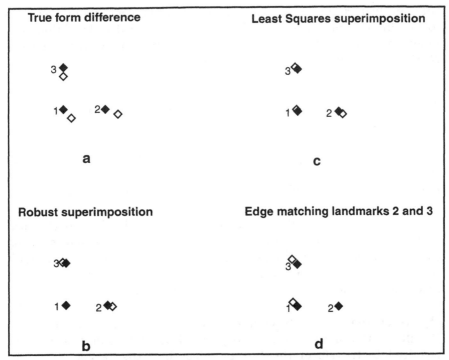

Figure 4.2. Superimposition of two triangles (three landmarks) according to various schemes. In 4.2a the "true form difference" that we created is given. Fig 4.2b shows the form difference depicted by the robust superimposition method. Fig. 4.2c shows the form difference depicted by the least squares method. Figure 4.2d provides the form difference depicted by edge superimposition. None of the superimposition methods depicted recreate the "true form difference." The various superimposition schemes provide varying versions of the overall comparison with dissimilar descriptions of the local differences between the two original triangles.

cies (or any forms) can never be known. The original relationship between the transparencies cannot be known because there is *no single common coordinate system* in which to put them. We cannot, therefore, reproduce the original relationship. Of course, they can be superimposed using any system of choice, but adoption of a particular superimposition scheme determines the relationship that will be "discovered." Comparison of the two forms using one superimposition scheme will produce one relationship, while comparison of the same two forms using another superimposition scheme will produce another relationship. No information is available that enables identification of which relationship is the true one. For the scientist, this means that this method cannot give reliable information concerning how two forms differ from one another. The mathematical details of this non-identifiability are provided in Part 2 of this chapter (see also Lele and McCulloch, 2000).

4.5 Transformational grids and deformation-based approaches

Deformation approaches, sometimes including the use of transformational grids, provide an alternate way to study form difference. These methods present the difference between forms as the changes required to transform one object into the other. Differences are expressed as magnitudes along specified directions and are depicted by the way in which a uniform grid placed over the original object changes or *deforms* during the transformation from the initial to the target object. However, as with the superimposition approach, the loss of a common coordinate system has profound effects for the description of form change.

In this discussion of deformation approaches, we refer to one of the objects as the initial object and to the other as the target object. Results of the analysis will be expressed as those changes that are required to transform the initial object into the target object. To explain the deformation approach, consider a lump of bread dough. Flatten the top of the dough ball and draw a square grid directly onto it. Now, plot the coordinates for the first object (e.g., the three landmarks of the red transparency) directly onto the grid that is drawn on the dough. Next, use your hands or a rolling pin to push, pull, or roll the dough in any manner you please. The relative positions of the landmarks have changed, as well as the appearance of the grid. The distorted grid is an excellent descriptor of the form change. It "localizes" and summarizes the form change/difference relative to the original, uniform grid.

Now take a blank, circular transparency and place it over the rolled dough. Trace the changed landmark positions onto this transparency using a green marker. This landmark configuration becomes the deformed or target object in this experiment. Refer to it as the green transparency. To simulate the loss of the common coordinate system, do two things: 1) roll the flattened dough with the grid markings back into a ball, and 2) shuffle the red and the green transparencies. We need to determine if it is possible to recreate the (distorted) grid that was on the dough given these two transparencies representing the initial and target configurations.

Transformation grids are thought to recreate the true form change grid as our bread dough did in the above experiment, but there are several reasons why they fail to do so. These reasons, some of which were introduced at the beginning of this chapter, are discussed with specific reference to deformation approaches below.

Arbitrary choice of deformation

There are many ways in which one object can be deformed into any other given object. Knowledge of material properties may help us to expect certain classes of deformations, but in most morphometric applications, material properties are neither available nor are explicitly considered, and the choice of deformation is arbitrary.

Sir D'Arcy Thompson recognized that the combined action of many different appropriate forces on any material form could transform one form into another. Thompson suggested that only simple transformations should be considered in biology, citing the laws of physics to support the use of parsimony in choosing a deformation. Parsimony is a guiding principle in many branches of biology, and in the case of deformation approaches, minimum energy deformations correspond to a specific type of parsimony. But without a valid scientific reason for the adoption of this principle in any particular study, the use of parsimony is difficult to defend. Parsimony is often defended on the basis of statistical reasons or mathematical convenience, but these reasons are a poor substitute for biological relevance.

Relationship between the number of landmarks and the choice of deformations under parsimony

Recall that in our example we have only three landmarks on each of the objects. With only three landmarks on each object, the simplest deformation that deforms one object into another is the affine deformation. Goodall and Green (1986) provide an excellent description of affine deformation. Simply stated, an affine transformation is a transformation such that lines that are parallel in the original grid remain parallel in the deformed grid. Equivalently, a circle drawn on the original grid is transformed into an ellipse in the transformed grid. Thus, shear (change on an axis not parallel with the original grid) does not occur in the affine transformation.

Think about the bread dough that you rolled out in the last experiment, and you will realize that the deformation resulting from rolling the dough may not necessarily be affine. That is, shear may have actually occurred, but because we have only three landmarks the best we can infer under the maxim of parsimony is that the deformation is affine. The mathematical reason for this situation is very similar to that found in classical regression. If only two data points are available, a straight line is the only curve that can be fit to the data. No matter how nonlinear the true relationship between the covariate and the response variable might be, the regression model can only provide a linear relationship because there are only two data points. A more informed idea about the underlying nonlinear relationship requires additional data points. For example, if the underlying relationship is quadratic, a minimum of three data points is required. Similar to this situation, in the study of deformation of objects, unless one has a substantial number of landmarks, the type of deformations that may be inferred is limited and is related to the number of landmarks available for study. This means that the deformation representing the most parsimonious solution changes with the number of landmarks. It can be shown that addition or subtraction of just one landmark changes the estimate of form change. This indicates that the original estimate cannot equal the true form change.

Lack of a common coordinate system

Suppose that the underlying deformation is affine and that three landmarks will, in fact, provide adequate information for determining this deformation. Can the information within the red and green transparencies that have been disturbed from their original positions provide this deformation? The same difficulty faced in the superimposition case presents itself here: we do not know and can never know the configuration of landmarks (the red and green transparencies) in their original version that includes information on their relative position in space. All that we have are the red and green transparencies which are rotated and translated versions of the original landmark configurations. Since we do not know the relationship of the original configurations to these translated and rotated versions, we can never know "the" coordinate system within which to conduct the deformation study.

To clarify, let's go back to the lump of bread dough and assume that the underlying deformation caused by the rolling of the dough is, in fact, affine. The affine deformation can be described using the following notation. Let M_1 and M_2 denote the landmark coordinate matrices corresponding to the two objects in their original positions. These matrices consist of the coordinates of the landmarks drawn on the transparencies before they are disturbed or thrown onto the floor. These are 3×2 matrices with each row representing the two-dimensional coordinates of the corresponding landmark. Under the assumption that the transformation from M_1 to M_2 is affine, we find this transformation by solving the equation: $M_2 = M_1A + \underline{1}t$ where A is a 2×2 matrix corresponding to the affine transformation (degree of change parallel to the initial grid), t is a 1×2 vector corresponding to the translation required to go from M_1 to M_2, and 1 is a 3×1 matrix of 1's. This is a linear system of six equations and six unknowns with a unique solution. The matrix A fully describes the deformation from M_1 to M_2. However, we have no knowledge of M_1 and M_2 in their original versions or positions. All that we know are the rotated and translated versions of M_1 and M_2, namely, M_1^* and M_2^*, where $M_1^* = M_1R_1 + \underline{1}t_1$ and $M_2^* = M_2R_2 + \underline{1}t_2$, R_1 and R_2 denote the unknown rotation parameters (R_1, R_2 are orthogonal matrices), and t_1 and t_2 are the unknown translation parameters.

The question then becomes, can we obtain the full and correct description of the true affine deformation by working with M_1^* and M_2^*? That is, if we solve the equation: $M_2^* = M_1^*A^* + \underline{1}t^*$, can we obtain the full and correct description of the true affine deformation? The answer is yes if, and only if, $A^* = A$. But a simple mathematical fact is that A^* does not equal A. Although the singular values of A^* and A are equal, the left and right eigenvectors are different due to the rotation matrices (R_1, R_2). The lack of a common coordinate system precludes us from finding full information, which includes not just the eigenvalues but also the eigenvectors, corresponding to the affine deformation.

4.6 The relationship between mathematical and scientific invariance

We have shown that the number of landmarks and the lack of a common coordinate system limit what can be learned about the true form change from the deformation-based description of form change. The superimposition approach cannot provide the vectors that depict the true form change vectors. Similarly, transformation grids cannot recreate the true deformation. Both approaches fail due to the nonidentifiabilty of the true common coordinate system in which forms can be compared. Any coordinate system can be chosen, but that choice has implications for statistical results and scientific interpretations. Only those inferences that are invariant to such an arbitrary choice are acceptable.

There is an exception to the strict application of this principle. The application of the principle of invariance may be relaxed if, under *all possible* choices of the external constraints in the superimposition approach or the deformation approach, the scientific conclusions drawn are identical (or at least similar enough to make no practical difference). In this case, the scientific results will be effectively invariant to the external constraints. To determine whether noninvariant descriptors of form change based on superimposition and deformation produce "scientifically invariant" answers, *all* noninvariant descriptors must be applied. This requires a great deal of work. Beyond this consideration, the following simple example is given to show that scientific invariance does not necessarily hold in real biological situations.

The method of Roentgenographic cephalometry (RCM) was introduced earlier in this chapter. A cephalometric radiograph is an x-ray produced under controlled conditions that allows for correction of distortion and enlargement, both of which occur during exposure of the film, by an x-ray beam. The data analyzed can consist of two-dimensional coordinates of landmarks identified on the cephalometric radiographs. Clinicians and dentists compare x-rays of skulls using this method. The method is often applied to compare the x-ray of a normal child with that of a child with a disorder, or to compare x-rays of a child taken at different points during growth (see Figure 4.1). The goal is to be able to describe and quantify differences between forms or samples of forms with reference to specific biological landmarks.

Apert syndrome is a genetic disorder that affects the bones of the skull. The genetic mutations responsible for this syndrome are known (Reardon and Winter et al., 1994; Park and Meyers et al., 1995; Wilkie and Slaney et al., 1995; Park and Theda et al., 1995c; Oldridge and Zackie et al., 1999). A method like RCM is applied to x-ray or computed tomography data to understand the craniofacial phenotypic aspects of this syndrome for the purpose of planning possible surgical procedures, planning the timing of these procedures, and inferring the biological processes that are responsible for the obvious difference in craniofacial form.

Let us again take the tracings of two cephalometric radiographs, one from a normal child at 2½ years of age (broken line in Figure 4.3) and one of a 2½-year-old child with Apert syndrome (solid line in Figure 4.3). Following traditional procedures, we locate sella in the center of the pituitary fossa and nasion at the intersection of the nasal and the frontal bones. Next, we match the two tracings by registering on the landmark sella (matching its location exactly in the two configurations) and rotating the two tracings until the lines that stretch from sella to nasion overlay one another. This is the superimposition seen in Figure 4.3a. Now take the same two tracings and register on sella, but superimpose on a line stretching from sella to the landmark basion (the most anterior point on the rim of foramen magnum). This superimposition is shown in Figure 4.3b. It is obvious that the two superimposition schemes produce varying summaries of the differences between the two forms. The superimposition scheme on the top shows the face of the Apert individual elevated and projected posteriorly as compared to normal, while that on the bottom shows the Apert face projected superiorly and posteriorly but to a different degree and in a different direction. The posterior portions of the calvaria match fairly well in the comparison at the bottom, but show marked differences using the superimposition shown at top. Since decisions regarding pathological processes and clinical treatment are made on the basis of these types of comparisons, it is critical to understand that the results vary depending upon (are not invariant to) the superimposition scheme used. This example is not unlike superimposition methods currently being applied in neuroscience, where specific features of the brain are mapped by registering on a "model" brain or atlas.

To underscore the influence of the choice of a superimposition scheme in the context of the analysis of biomedical data, we further scrutinize the example given above. Eleven two-dimensional landmarks were located on the cephalometric radiographs of a sample of children diagnosed with Apert syndrome from the Center for Craniofacial Anomalies, University of Illinois, Chicago, and on the radiographs of unaffected children of similar age from the Bolton-Brush Growth Series, Cleveland, OH. The landmark data were originally presented in Richtsmeier (1985). Figure 4.4a shows the placement of the landmarks on a radiographic tracing (upper left quadrant) and comparison of the estimated mean forms for the two samples using various superimposition schemes (Figures 4.4b,c,d). Landmarks corresponding to the unaffected sample mean form are shown as solid diamonds, whereas the open diamonds correspond to the landmark locations on the mean form of the Apert sample. The superimposition shown in Figure 4.3b

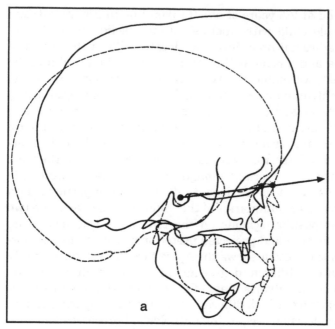

a

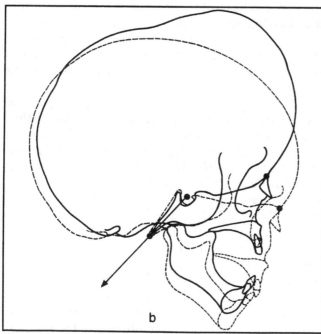

b

Figure 4.3 The superimposition of radiographs of similarly aged children, one diagnosed with Apert syndrome, the other unaffected. Figure 4.3a shows a comparison of the two forms based on a sella-nasion superimposition scheme. The comparison shown in Figure 4.3b is based on a sella-basion superimposition scheme. Differences in the two comparisons are obvious. The choice of which superimposition scheme is used might ultimately influence the planning of patient treatment (following Richtsmeier, 1986).

mimics the standard in RCM comparisons where the position of land-mark 7 (tuberculum sella) is matched exactly on the two forms (samples of forms) and the forms are oriented along a line stretching from landmark 7 to landmark 1 (nasion). The cranial base is common-ly used for superimposition as described above because it is thought to be biologically stable, but this assumption of stability affects the results of any comparison. Under this superimposition scheme (Figure 4.4b), we conclude that there is no difference local to tuberculum sella, but that the posterior neurocranium (represented by landmarks 10 and 11) in the Apert skull is shallower than the normal skull. The face

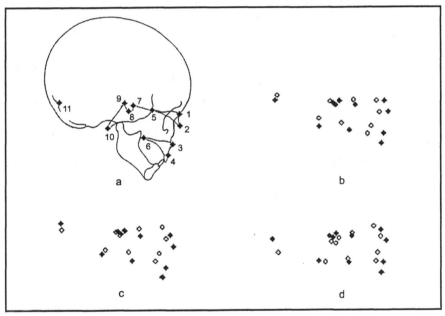

Figure 4.4 This figure demonstrates the lack of invariance of superimposition approach-es. The upper left quadrant of this figure provides the placement of the landmarks used in the analysis as closed diamonds on a tracing of a radiograph of an unaffected child's skull. In the other three quadrants, the position of corresponding landmarks located on the esti-mated mean forms of the Apert syndrome sample are shown as open diamonds, while the locations of the corresponding landmarks on the mean form of the unaffected sample are shown as closed diamonds. The upper right quadrant (b) shows the results of comparing data collected from the Apert and unaffected samples using the sella-nasion superimposi-tion. The lower left quadrant (c) shows the results of comparing the same data using the sella-basion superimposition. The lower right quadrant (d) shows the results of comparing the same data using the least squares superimposition. Notice that the local differences between the two forms vary according to which superimposition scheme is used.

(landmarks 1, 2, 3, and 4) does not extend as far forward in the Apert individual. The upper face (landmarks 1 and 2) is shifted posteriorly, whereas the lower face and palate (landmarks 3, 4, and 6) are shifted posteriorly and superiorly.

In Figure 4.4c the same two mean forms are compared using another superimposition scheme that matches landmark 7 (tuberculum sella) exactly and orients the forms on a line stretching from landmark 7 to landmark 10 (basion). In this comparison, similar to the previous superimposition scheme, there is no difference in the forms local to the sella region (landmarks 7, 8, and 9). However, landmark 11 is displaced in the Apert individual, in a pattern that is opposite to what we obtained by the first superimposition. All facial landmarks of the Apert individual (landmarks 1, 2, 3, 4, and 6) are displaced posteriorly and superiorly.

Finally, in Figure 4.4d, we show the superimposition obtained according to the least squares criterion. This provides another quite distinct description of how the two mean forms differ from each other. In this comparison, landmarks local to the pituitary fossa (landmarks 7, 8, and 9) on the Apert cranial base are displaced inferiorly. The anterior face (landmarks 1, 2, 3, and 4) of the Apert individual is displaced superoposteriorly. There is minimal difference local to the posterior portion of the palate (landmark 6). The posterior neurocranium of the Apert individual is displaced inferiorly, indicating a shallower neurocranium in the unaffected individual.

Now, let us remind ourselves of the reasons for comparing these forms. One reason involved providing the surgeon with information on how the Apert syndrome skull differs from the normal skull. A surgeon might design varying corrective procedures depending upon which superimposition scheme is applied to the data. It should be clear from this example that the application of the superimposition approach provides very different scientific inferences depending on the choice of the external constraint.

Now consider the comparison of the same two mean forms using the thin-plate spline methodology. One of the attractions of the thin-plate spline methodology is the ability to present the form change graphically as the way in which a square grid changes under the estimated deformation. In Figure 4.5 we show the deformation grid that results from the comparison of the normal and Apert data when the constraint is based on the minimum bending energy thin-plate spline (Figure 4.5b), and when the constraint is based on $W = D^{-1}$ where D is the matrix of squared distances between all pairs of landmarks in the normal child (Figure 4.5c). It is clear that the two grids are quite different, leading to different scientific conclusions about the locus, magnitude and direction of the differences between the two forms. Looking at the inverse-squared distance weighted grid (Figure 4.5b), we see that the grid has been markedly compressed on the lower left corner. This represents compression of both the posterior aspect of the neurocranium (landmarks 10 and 11) and the distance from the back of the palate (landmark 6) to the anterior aspect of the foramen magnum (landmark 10). A less obvious constriction is noted on the upper right corner of the grid while the lower right corner is stretched. This indicates constriction of the upper face (landmark 1) and stretching of the lower face (landmarks 3, 4, and 6). Within the grid, anteroposterior expansion or stretching is shown local to the pituitary fossa region of the cranial base (landmarks 7, 8, 9). The external shape of the minimum bending energy grid shows comparatively less deformation (Figure 4.5c). The face (landmarks 1, 2, and 4) shows some superoinferior constriction, while additional restriction occurs along the

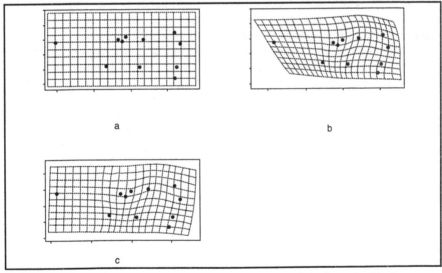

Figure 4.5 The noninvariance of the deformation approach is shown in this figure. Part 4.5a provides the initial grid with landmarks in place on the estimated mean form of the normal sample. Panel 4.5b shows deformation of the normal mean form into the mean form estimated for the Apert sample based on the inverse squared distance weighted deformation. Panel 4.5c gives the same deformation but this time based on minimum bending energy. The deformation grids produced from the comparison of the same data sets differ according to the choice of the side condition. Scientific conclusions of the two deformation analyses differ.

anteroposterior axis in the lower center of the grid (between landmarks 6 and 10). The pituitary fossa (landmarks 7, 8, 9) shows stretching along both the superoinferior and anteroposterior axes.

The form difference measure based on the ratio of the singular values of the affine deformation parameter used in the two thin-plate spline approaches also differs. In the case of minimum bending energy the value is 1.1 (0.897/0.807), and in the case of the inverse square distance-weighting scheme it is 2.6 (1.731/0.665). These are obviously different quantities and affect our interpretation of the deformation analysis. In the case of the minimum bending energy constraint, the affine component of the form difference is relatively small, and we assume that the difference between the forms could be explained by the nonaffine deformation. On the other hand, according to the inverse-squared distance constraint, there is a substantial affine component. This analysis of real data demonstrates that statistical and scientific inferences can change considerably depending on the choice of the constraint. This example illustrates that the noninvariance of the superimposition and deformation approaches is not simply a mathematical detail. Ignoring the principle of invariance can have significant implications in practical and scientific terms.

In conclusion, we repeat that the number of landmarks and the lack of a common coordinate system limit what can be learned about the true form change from any deformation-based description of form change. We have shown that superimposition approaches cannot provide the vectors that depict the true form change vectors, nor can transformational grids recreate the true deformation. Both approaches fail due to the nonidentifiabilty of the true common coordinate system in which forms can be compared. Any coordinate system can be chosen for use in analysis, but that choice has implications for statistical results and scientific interpretations. Only those inferences that are invariant to such an arbitrary choice are acceptable. We turn now to the introduction of an alternate approach to the comparison of forms that satisfies the invariance requirement.

4.7 An invariant approach: Euclidean Distance Matrix Analysis (EDMA)

We have just shown that nuisance parameters can affect results in unpredictable ways. We have shown how the arbitrary choice of a coordinate system for analytical purposes has unwanted effects on empirical results and scientific conclusions. The alternative approach is to choose not to adopt a coordinate system. In this section, we present an approach that enables the comparison of forms independent of any and all coordinate systems. This is the approach that we favor. The following section clarifies the significance of comparing forms using a coordinate system-free method. We begin by presenting the concept of "orbit" as it relates to the study of form difference. Understanding this concept will reinforce our insistence on using only that information that we know unambiguously when comparing forms.

4.7.1 Orbits of equivalent forms and description of form difference

Let M be a landmark coordinate matrix of dimension $K \times D$. Following our definition, the form of an object as represented by this collection of landmark coordinates is that characteristic which remains invariant under the group of transformations consisting of rotation, translation, and reflection. Now, think of the collection of *all* $K \times D$ matrices that can be obtained by any rotation, reflection, and translation of M. That collection of matrices is called an *orbit* described by M under the group of transformations consisting of rotation, reflection, and translation (Figure 4.6). All matrices within any single orbit represent exactly the same form because they differ only on the basis of translation, rotation, or reflection. An invariant descriptor of form change describes the difference in the orbits that the forms occupy.

As an aid to understanding the idea of an orbit, consider a topographic map of a mountainous area. Each contour on a topographic map corresponds to a surface of constant (equal) elevation. No matter where a point lies in longitude or latitude, elevation remains the same as long as the point stays on the defined contour. In other words, as long as a point remains on a single contour line, elevation is *invariant* with respect to latitude and longitude. Orbits in the space of all form matrices as defined above are similar to contours on a topographic map.

Suppose that a person is hiking on the surface described by the topographic map. Suppose further that this hiker carries a sensor that sends out a signal that identifies only his elevation exactly. If we are trying to locate this hiker but only have the signal from his sensor, we can place that hiker onto a contour defined by a particular elevation, but not to any particular location on that contour. This is analogous to the limitations of our knowledge when dealing with landmark coordinate data. All that we can know with certainty is the orbit to which the

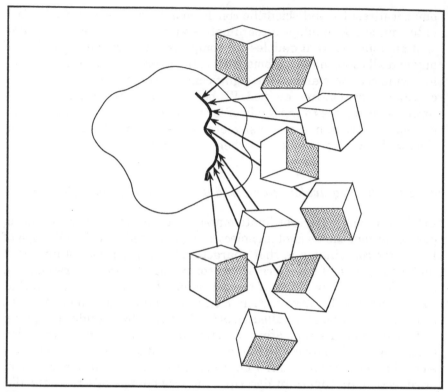

Figure 4.6. Graphic representation of an orbit for a cube of specific dimensions with eight landmarks. All rotated, reflected, and translated versions of the 8x3 matrix representing this cube map to the orbit depicted as a curve.

form under study belongs in the space of all $K \times D$ matrices. We cannot know the exact location of the original $K \times D$ matrix on the identified orbit.

Now, suppose we know that the hiker started his hike at elevation of 2000ft. above sea level but he is now at an elevation of 2560ft. We know the contour on which the hiker started and the contour that he presently occupies, but we have no information regarding his exact starting place, nor his exact location at the current elevation. This is analogous to the problem of form comparison. We know the orbit of the first form and the orbit of the second form, but we do not know the exact location of the forms on their respective orbits.

If our information is limited to elevation, we can only know that the hiker has ascended 560 vertical feet. This finding is invariant to the exact location of the hiker on the initial contour and invariant to the final location of the hiker on a given contour. Although we would like to describe the exact path that the climber followed to gain this elevation, without further information we cannot. We can assign a "sensible" starting point for the hiker, but this information may be wrong and eventually misleading. The description of the exact path can be provided if, and only if, we know exactly the point where the hiker started and every intermediate point that he traversed as he attained the higher elevation. If our information is limited to elevation, we simply do not have the information to obtain a description of his path.

To take this analogy a bit further, let us say that we know the exact position of a specific path between the elevations of 2000ft. and 2560ft. The most parsimonious conclusion in this case is that the hiker used this path to gain 560ft. in elevation. But, perhaps the path was designed for beginners and this particular hiker prefers a challenge. The hiker forges trails up embankments and scales walls to reach 2560ft. We do not have this information, however, and parsimony forces us to assume that the hiker used the well-known path. In this case, the assumption of parsimony provides us with incorrect information regarding how the hiker got from his starting point to his ending point. As scientists, our choice is between using only the information that is known, or including information or assumptions that are not testable but that provide a more "complete" answer.

The complete mathematical description of these concepts is presented in Part 2 of this chapter. A simple and brief mathematical description is presented here. Let M_1 be any point on the first orbit in the space of the landmark coordinate matrices. The orbit of M_1 contains all translated, rotated, and reflected versions of M_1. Let M_2, which is also a two-dimensional three-landmark object, be represented by any point *on the second orbit* in the space of the landmark coordinate matrices. Let $Diff(M_1M_2)$ denote a description of the difference between the two forms. Because we only know the orbits on which the two forms lie but not the exact locations of the forms on the orbits, the description of the difference between the two forms should not require information pertaining to the exact locations on the orbits. Instead this description should be invariant to the exact location of the forms on their respective orbits (Lele and McCulloch, 2000). Mathematically, this is written as $Diff(M_1, M_2) = Diff(M_1R_1 + \underline{1}t_1, M_2R_2 + \underline{1}t_2)$.

Inclusion of information other than what is unambiguously known may provide a seemingly more complete, but potentially misleading answer. The information that is unambiguously known consists of identification of the orbit on which the form lies. An invariant descriptor of form change utilizes only this unambiguous information and describes the difference in orbits that the forms occupy. We now use the principles of EDMA as presented in Chapter 3 to create descriptors of form change that are invariant and useful to the investigator who is striving to propose scientific inference from observations of form.

4.7.2 The form space

We continue with a simple situation in which we have a form represented by three landmarks. Recall that the form of an object represented by three landmarks can be expressed as a vector of three distances. We have shown that the vector of three distances is an invariant representation of the form. That is, no matter how the object is rotated, reflected, or translated, the vector of three distances remains unchanged.

Let us consider the object represented by the landmark coordinate

matrix $M = \begin{bmatrix} 0 & 0 \\ 1 & 0 \\ 0 & 1 \end{bmatrix}$.

We have seen in the last chapter that the form matrix corresponding to this object is given by

$FM(M) = \begin{bmatrix} 0 & 1 & 1 \\ 1 & 0 & 1.41 \\ 1 & 1.41 & 0 \end{bmatrix}$, and that the form matrix can be written

as a vector of all entries above the diagonal, $\begin{bmatrix} 1 \\ 1 \\ 1.41 \end{bmatrix}$.

Now consider all distinct *three-landmark objects that form a triangle*. There is a vector of three distances corresponding to each of these objects. Each vector, although consisting of linear distances, corresponds to a point in a three-dimensional Euclidean space. The collection of points corresponding to *all* three-landmark objects is called the *form space* of three landmark objects. Figure 4.7 shows the form space of all three-landmark objects that form triangles with vectors corresponding to the measure of each linear distance of the triangle. The vector of three distances is recorded as a single point in

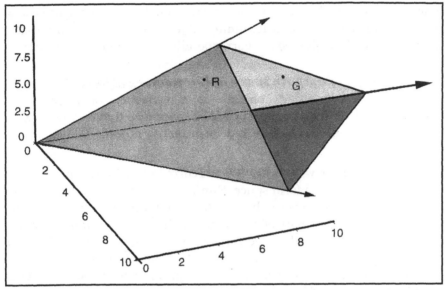

Figure 4.7. The subset of a three-dimensional Euclidean space that corresponds to the form space of three-landmark objects. Every point in this subset corresponds to some three-landmark object, and every three-landmark object corresponds to some point in this subset. Points labeled R and G represent the location of the red and green transparencies (three-landmark forms) in this space.

this space.

There are certain constraints on three-landmark objects that form triangles that make this space only a *subset* of the entire three-dimensional Euclidean space. For example, dimensions measured on an object must be positive, and so any distance corresponding to the side of a triangle on this object must be positive. Consequently, the location of three-landmark objects in this form space will be restricted to the positive quadrant of Euclidean space. In addition, a set of three positive numbers corresponds to a triangle if and only if the sum of any two numbers exceeds the third. This further restricts the subset of space corresponding to three-landmark objects that form triangles. There are parts of the Euclidean space where this constraint is not met and therefore cannot be part of the defined form space.

Generalization of these findings to K-landmarks is fairly straightforward. We have seen in Chapter 3 that the form of any K-landmark object, whether two- or three-dimensional, can be represented by the collection of all possible pair-wise distances. There are $L = K(K-1)/2$ such distances, so that the dimension of this space is equal to L. Thus, one can plot a single point in the L-dimensional Euclidean space corresponding to each distinct K-landmark object. The collection of all such points is called the *form space* corresponding to K-landmark objects.

> ***Definition of the form space of K-landmark objects:*** The form space of K-landmark objects is a collection of points in $L = K(K-1)/2$ dimensional Euclidean space such that each point in this collection corresponds to a form matrix of some K-landmark object.

Every possible form with K-landmarks corresponds to one and only one point in the identified form space. Similar to the three-landmark case, this form space occupies only a subset of L-dimensional Euclidean space (e.g., only that part of the space with positive numbers). When the forms being studied include more than three landmarks (K > 3), the additional constraints that specify the form space subset are complicated. The exact mathematical description of these constraints is provided in Part 2 of this chapter.

4.7.3 Studying the difference between two forms using the form space

Suppose we have two objects. Any two objects will do, but let's think back to the red and green transparencies described previously. Given our definition of form space, there is a single point in the form space corresponding to the red transparency and all other three-landmark objects with the same form. Similarly, there is another single point in the form space corresponding to the green transparency and all other objects with the same form. Since we are dealing with three landmark objects, we can plot them in the form space as points R and G in Figure 4.7. How can the difference between points R and G be determined?

There are many different ways to describe the difference between points R and G. We will present a few possibilities that have proved useful in our own research. As long as this description depends solely on the location of points R and G in the defined form space, and not on any other extraneous information, the description will satisfy the invariance requirement. Definitions of form difference should satisfy the invariance requirement and be biologically interpretable.

Defining the difference between two forms

Let us begin with two forms, A and B. We will set the landmark coordinates for forms A and B to differ only in the position of landmark 2, so that the coordinate locations for these two forms are:

$$A = \begin{bmatrix} 0 & 0 \\ 1 & 0 \\ 0 & 1 \end{bmatrix} \qquad B = \begin{bmatrix} 0 & 0 \\ 2 & 0 \\ 0 & 1 \end{bmatrix}$$

FM (A) and *FM (B)* denote the form matrices corresponding to the two objects. Recall that FM_{ij} *(A)* is the distance between landmarks i and j in object A and that FM_{ij} *(B)* is the corresponding distance in object B. The form matrices for the above objects are:

$$FM(A) = \begin{bmatrix} 0 & 1 & 1 \\ 1 & 0 & 1.41 \\ 1 & 1.41 & 0 \end{bmatrix} \qquad FM(B) = \begin{bmatrix} 0 & 2 & 1 \\ 2 & 0 & 2.23 \\ 1 & 2.23 & 0 \end{bmatrix}$$

We present three possible ways to express form difference using the form matrices, FM (A) and FM (B).

Arithmetic form difference

The difference between forms can be expressed as the arithmetic difference between like linear distances in the two forms. The collection of these differences can be expressed as a matrix consisting of elements that correspond to the arithmetic difference between like linear distances in the two objects. For example, the absolute form difference matrix, *AFDM (B, A)* = FM_{ij} *(B)* − FM_{ij} *(A)*, for the above two objects, is given as:

$$AFDM(B, A) = \begin{bmatrix} 0 & 1 & 0 \\ 1 & 0 & 0.82 \\ 0 & 0.82 & 0 \end{bmatrix}$$

Relative form difference

The relative difference between forms is another useful way of describing form difference in biological objects. When written as a matrix, the collection of relative differences is called the form difference matrix. Our emphasis on this particular expression of form difference has resulted in the following notation, $FDM\ (B,A) = \begin{bmatrix} \dfrac{FM_{ij}(B)}{FM_{ij}(A)} \end{bmatrix}$, where elements of the form difference matrix correspond to the ratios of like-linear distances, and the division is done element-wise (where $0/0 = 0$ by definition). Continuing with the data sets described above, the form difference matrix for forms A and B is written:

$$FDM(B,A) = \begin{bmatrix} 0 & 2 & 1 \\ 2 & 0 & 1.59 \\ 1 & 1.59 & 0 \end{bmatrix}$$

Because form difference matrices are square, symmetric matrices with zero along the diagonal, only the above diagonal elements are needed to describe the form difference. As a matter of convention, we write the form difference matrix in vector form, where only the above-diagonal elements are reported:

$$FDM(B,A) = \begin{bmatrix} 2 \\ 1 \\ 1.59 \end{bmatrix}$$

Scaling factors and differences in scaled forms

In many biological studies, it is helpful to adjust for size differences according to some scaling factor before comparison. Within the context of landmark coordinate data, one can use various measures as scaling factors. For example, a specific biological distance might be used, or the geometric mean of all distances could be used as a scaling factor. Mosimann (1979) provided a general mathematical definition of scaling factor and referred to scaling factors as a measure of 'size.' Following these ideas, the scaled form is referred to as 'size-corrected.' Comparison of size-corrected forms is often considered a comparison of 'shapes.'

Following Mosimann (1979), we consider any function of the distances that always takes a positive value as a scaling factor. However, the terminology of 'size' and 'shape' that follows from the use of scaling factors is not precise (Mosimann, 1979; Lele, 1991). Surrogate measures for 'size' are many (e.g., volume, weight, length) and the use of each will result in a different size-corrected shape. Keeping this imprecision in mind, we adopt the common usage of the terms 'size' and 'shape' in the following discussion.

Suppose we calculate the geometric mean of all distances in $FM(A)$ and refer to the geometric mean as 'size.' Let $S(A) = \{\prod FM_{ij}(A)\}^{1/L}$ denote the 'size' of object A. Using the ongoing example, the geometric mean or size of form A is 1.12. Similarly, let $S(B)$ denote the 'size' of form B. In the example above, $S(B) = 1.64$. Now consider the matrix $SM_{ij}(A) = \dfrac{FM_{ij}(A)}{S(A)}$ We consider $SM_{ij}(A)$ to represent the 'scaled form of A' or the *shape matrix* of A. Similarly, we define $SM_{ij}(B) = \dfrac{FM_{ij}(B)}{S(B)}$ as the shape matrix of B. These matrices are given as:

$$SM(A) = \begin{bmatrix} 0 & 0.89 & 0.89 \\ 0.89 & 0 & 1.26 \\ 0.89 & 1.26 & 0 \end{bmatrix} \text{ and } SM(B) = \begin{bmatrix} 0 & 1.21 & 0.61 \\ 1.21 & 0 & 1.36 \\ 0.61 & 1.36 & 0 \end{bmatrix}$$

One can describe the form difference between objects A and B by using two quantities. $\dfrac{S(B)}{S(A)} = \dfrac{1.647}{1.122}$ quantifies the 'size difference' between forms A and B, while the shape difference between forms is given by a shape difference matrix whose elements represent the absolute difference between like-linear distances of the two shape matrices, $SDM_{ij}(B,A) = SM_{ij}(B) - SM_{ij}(A)$. In our example, the shape difference matrix is:

$$SDM_{ij}(B,A) = \begin{bmatrix} 0 & 0.32 & -0.284 \\ 0.32 & 0 & 0.097 \\ -0.284 & 0.097 & 0 \end{bmatrix}$$

4.7.4 Estimation of form difference using EDMA

Up to this point, we have been discussing the comparison of single forms and have operated under the assumption that the true means are available. Remember, however, that in practice the true means are never available. All that are available are rotated and translated versions of the mean. We have demonstrated that because of the nuisance parameters of translation, rotation, and reflection, only the form matrix corresponding to the coordinates of the mean form can be estimated.

In most research situations, we must estimate the mean template from a sample of available observations. We will denote the *estimated* form matrix corresponding to the mean of the first group by $FM(\hat{A})$ and the *estimated* form matrix corresponding to the mean of the second group by $FM(\hat{B})$. The reader can think of an estimated mean form as a form matrix representing the average of the linear distances from all forms in the sample, although the exact estimation procedure is a bit more complex (see Part 2 of this chapter). The mean form difference matrix provides the estimation of the difference between mean forms. Details of the estimation procedures are given in Part 2 of this chapter.

Estimating the mean form difference matrix

The *estimated* mean form difference matrix is given by

$$FDM(\hat{B}, \hat{A}) = \left[\frac{FM_{ij}(\hat{B})}{FM_{ij}(\hat{A})} \right].$$

Estimating the scaling factor

Although there are a number of possible estimates that can be used as a scaling factor, in this example we use the geometric mean of the distances. In this case, the *estimated* scaling factor is given by $S(\hat{A}) = \{ \prod FM_{ij}(\hat{A}) \}^{1/L}.$

Estimating the mean shape matrix

The *estimated* mean shape matrix is given by $SM_{ij}(\hat{A}) = \dfrac{FM_{ij}(\hat{A})}{S(\hat{A})}$.

We emphasize that the definition of a shape matrix is dependent upon the choice of the scaling factor.

Estimating the mean shape difference matrix

The *estimated* mean shape difference matrix is given by

$$SDM_{ij}(\hat{B}, \hat{A}) = SM_{ij}(\hat{B}) - SM_{ij}(\hat{A})$$

Using ideas of the form space, we have provided ways to *estimate* form and shape difference as arithmetic differences between forms, as relative differences between forms, and arithmetic differences between scaled forms. We now discuss statistical methods that can be used to test for differences in form and shape.

4.8 Statistical analysis of form and shape difference using EDMA

Traditionally, the first step in statistical analysis is the testing of a simple null hypothesis that the two forms or shapes are identical to each other. However, an excessive emphasis on such a test may prevent further exploration of the data and may even mislead researchers by obscuring important local similarities or differences in forms or shapes. For this reason, we advocate and emphasize the use of confidence intervals when comparing forms. We present both approaches to the statistical comparison of forms.

4.8.1 Confidence intervals for localized form differences

Suppose that the *estimated* form difference matrix determines that a particular distance is larger in object *B* as compared to object *A*. How do we know that this difference is significant in a statistical sense? Basic statistics tells us that if we have a very large number of observations, or if the biological variability is very small, we can put greater trust in the estimated form difference matrix than if we had a small sample size and/or large variability. It is important that any estimate of a scientifically relevant quantity carry with it a statement about its reliability. Point estimates do not provide any information about reliability of the estimate. An interval estimate incorporates reliability into the estimate and can therefore be a powerful statistical tool (Kowalski, 1972; Reichardt and Gollob, 1997). When an interval estimate has a certain probability of including the true value of the parameter, that interval is called a *confidence interval*. We provide a brief discussion at the end of this section concerning the interpretation of confidence intervals for form and shape difference. For the precise meaning of confidence intervals, we ask the reader to refer to a statistical text (e.g., Casella and Berger, 1990; see also Harlow, Mulaik, and Steiger, 1997). The subtleties of interpretation cannot be fully covered here but it is important that the user understand them.

Lele and Richtsmeier (1995) introduced the procedure for obtaining confidence intervals for elements of the form difference matrix. There are two different approaches available for the computation of confidence intervals for the form difference matrix. The first approach, called the Monte Carlo approach, is model-based and assumes that the underlying perturbation model is Gaussian. In addition to assuming Gaussian perturbation, the Monte Carlo approach makes direct use of the parameters estimated from the samples under study (i.e., mean, variance-covariance matrix). The second approach, called the Bootstrap approach, is less model-based as dependence on the assumption of the underlying Gaussian perturbation model is relaxed (Efron and Tibshirani, 1991). The Monte Carlo approach is preferable if one has reasonable confidence that the data being analyzed follow the underlying model as Monte Carlo confidence intervals can be sensitive to deviations from the underlying model (Huber, 1972). For this reason, if evidence suggests that the data do not follow the model, the Bootstrap approach might be more suitable. However, if only a small number of observations are available, the Bootstrap approach may prove problematic and it may be necessary to adopt the Monte Carlo approach. Both the Monte Carlo and Bootstrap methods for obtaining confidence intervals are described below. The computational algorithms for these approaches are presented in Part 2 of this chapter.

4.8.2 The Bootstrap method for obtaining confidence intervals

The following steps briefly describe the Bootstrap procedure. A proper computational algorithm is provided in Part 2 for any reader who wishes to program this procedure.

Let $A_1, A_2, A_3, ..., A_n$ and $B_1, B_2, B_3, ..., B_m$ be the landmark coordinate matrices for individuals from samples representing populations A and B. Suppose we have obtained a sample of size n from population A and a sample of size m from population B.

> STEP 1. Obtain a simple random sample with replacement of size n from sample $A_1, A_2, ..., A_n$ and of size m from sample $B_1, B_2, ..., B_m$. Let us refer to the samples from the first population as $A^*_1, A^*_2, ...A^*_n$ and those from the second population as $B^*_1, B^*_2, ...B^*_m$.
> STEP 2. Calculate the mean form difference matrix, $FDM(B^*, A^*)$ for the samples obtained in Step 1 using the equations given previously.

STEP 3. Repeat Steps 1 and 2 for C times where C is sufficiently large (i.e., from 200 to 1000).

Each $FDM(B,A)$ calculated from a Bootstrap sample is written as a
vector with $K(K-1)/2$ entries. The collection of all $FDM(B,A)$ obtained
in this way is written as a matrix that has $K(K-1)/2$ rows and C
columns. Each FDM is sorted according to the landmarks that define
the linear distance. Each column is a form difference matrix in vector
format obtained at the end of Step 2, and each row represents C form
difference ratios for a linear distance between a specified pair of landmarks. To obtain a confidence interval for each linear distance, the
ratios in each row are sorted in increasing order. The $100(1-\alpha)\%$ confidence interval is delimited by removing the first $\alpha/2$ % and the last $\alpha/2$
% of the sorted entries. The minimum and maximum entries remain-
ing in that row constitute the lower and upper confidence limits for
that particular linear distance. If the interval of entries spanning the
lower and upper confidence limits contains the value 1, then it is likely that the particular distance is not different in the two populations.
The interval also provides some idea as to the range of values that particular ratios might take. For example, if a confidence interval is
(1.3,1.7) we know that this particular linear distance is larger in sample B relative to sample A, but also that the linear distance in sample
B is likely to be 30% to 70% larger than the same distance in sample
A. Similar statements may be made separately for each row using the
values obtained to estimate the confidence interval for each linear distance.

4.8.3 The Monte Carlo method for obtaining confidence intervals.

As in the previous case, let $A_1, A_2, A_3,...A_n$ and $B_1, B_2, B_3 ..., B_m$ denote
the two samples.

STEP 1. Estimate the mean form and the variance-covariance
matrix for samples A and B.
STEP 2. Use estimates of the mean form and variance-covariance matrix based on sample A to generate a new sample of
observations (N=n) using the Gaussian perturbation model.
STEP 3. Use the estimates of the mean form and variance-
covariance matrix based on sample B to generate a new
sample of observations (N=m) using the Gaussian perturbation model.

STEP 4. Use the parametric samples generated in Steps 2 and 3 to calculate a new form difference matrix for these samples, $FDM(B^*, A^*)$.

STEP 5. Repeat Steps 2 through 4 for C times, where C is sufficiently large (i.e., from 200 to 1000).

Each $FDM(B^*, A^*)$ calculated from the Monte Carlo samples is written as a vector with $K(K-1)/2$ entries. A collection of Monte Carlo samples compared in this way and written as a matrix has $K(K-1)/2$ rows and C columns, each column being sorted according to the landmarks that define the linear distance. Each column is a form difference matrix in vector format obtained at the end of Step 4 and each row represents C form difference ratios for a linear distance between a specified pair of landmarks. Sort the values in each row in ascending order. The $100(1-\alpha)\%$ confidence interval is delimited by removing the first $\alpha/2$ % and the last $\alpha/2$ % of the sorted entries. If the interval of entries spanning the lower and upper confidence limits contains the value 1, then it is likely that the particular distance is not different in the two populations. As with the Bootstrap approach, this interval also provides a suggestion of the range of values that a particular ratio might take. Thus, if the confidence interval is say (0.85,0.95) we can conclude that the linear distance is smaller in sample B relative to sample A, and that it is likely that the linear distance is 5 to 15% smaller in sample B. Once a confidence interval is obtained for each linear distance, similar statements can be made separately using data from each row.

Notice that the only difference between the Bootstrap procedure and the Monte Carlo procedure is the way in which the samples in Step 2 are generated. The Bootstrap approach accomplishes this by using simple random sampling directly from the data, while the Monte Carlo procedure uses the Gaussian model to generate data from the estimated parameters.

An additional nuance of both the Bootstrap and Monte Carlo procedures is that the exact output of running either procedure will vary even if the input data remain the same. Suppose you run the Bootstrap program for obtaining confidence intervals at home on Monday evening. You leave the printout at home and have to re-run the Bootstrap confidence interval routine again the following morning at work. If you were to compare the two outputs that have used the exact same input data, they will be slightly different. The reason for this is that the samples obtained in Step 2 of both the Bootstrap and Monte Carlo procedures will not, in general, be the same from one analysis to the next. The full statistical explanation for this is beyond the scope of this monograph but a colloquial explanation should help here.

Both the Bootstrap approach and the Monte Carlo approach involve a randomized procedure. In the case of the Bootstrap approach, if we randomly pick individuals from the known sample, it is highly unlikely that the same random sample will be chosen upon repetition. This will result in slight differences in output when the whole procedure is run different times. The Monte Carlo approach is also a randomized procedure, but it is model-based. Even though the estimate of the mean and variance-covariance matrix is constant from analysis to analysis, the generation of data under the Gaussian model using these estimates will result in slightly different Monte Carlo samples each time the analysis is done. Consequently, the exact values of the limits of the confidence intervals will change as you repeat a confidence interval estimation analysis. The differences between analyses, however, are not so large as to affect your conclusions.

Let us try to understand the interpretation of the confidence intervals in an intuitive fashion. Suppose that there are 100 researchers in the world who are studying the difference in craniofacial form between unaffected children and children with Apert syndrome. Each scientist has a sample of patients and controls from his own clinic and collects the coordinates for the same landmarks from these individuals. When the affected and unaffected individuals from each clinic are compared using the above procedures, 90% confidence intervals are obtained. Each of the researchers uses the confidence intervals derived from their clinic data to make a decision regarding whether or not a particular linear distance is larger in Apert syndrome children as compared to unaffected children. Following the confidence interval interpretation, we say that approximately 90 of the 100 researchers are correct in this decision. Knowing this statistic does not guarantee that any particular researcher is correct about their decision, but on average, the community of researchers is right 90% of the time.

We provide percentile intervals because they are simple to interpret and are sufficient in most biological applications. We stress that the intervals provided are element-wise confidence intervals. This means that each confidence interval is computed for a specific linear distance and that they should be interpreted in that way. Finally, although the confidence intervals are reported separately for each linear distance (Lele and Richtsmeier, 1995), our computational procedures *do not* assume that the distances in the form matrix are independently distributed. Our procedures fully account for the dependence between the distances.

The confidence intervals described above can localize form difference to particular linear distances, and in some cases to particular landmarks. For that reason, they are more useful in revealing biological information and relationships than they are for testing equality of form or shape. The testing of global null hypotheses has a long history and is desired by certain scientists. For this reason, we present two different statistical procedures for testing the hypothesis of equality of shapes using point estimates.

4.9 Statistical hypothesis testing for shape difference

The testing of statistical hypotheses has become a standard and nearly required part of biological analyses. We deemphasize the importance of this procedure because it can distract the researcher from looking carefully at the data and direct the researcher away from data based inferences that are not directly revealed by the results of hypothesis testing. We include statistical hypothesis testing so as to satisfy the more traditionally minded biologists and to offer a complete suite of methods.

The rules for conducting a valid and rigorous statistical test of a null hypothesis are fairly standard in statistics (see Casella and Berger, 1990). Any statistical testing procedure has the following components: 1) a statement of the null and the alternative hypotheses; 2) a test statistic which takes a specified value if the null hypothesis is true and a different set of values under the alternative; and 3) the distribution of the test statistic when the null hypothesis is true. If the observed value of the test statistic lies in the extreme tails of the null distribution, the null hypothesis is rejected. One may also calculate the probability of getting the observed (or more extreme) value of the test statistic and report it as a *p-value*. Various approaches to hypothesis testing in the study of difference in form using EDMA are detailed below.

When comparing shapes using EDMA, the null hypothesis is that the mean form of one population is a scaled version of the other population (i.e., the ratios of all linear distances are equal to a constant). The inability to reject this null hypothesis would mean that the two samples of forms are similar in shape and differ only in size. The alternative hypothesis is that the two populations differ in form. In this case, the two populations are not merely scaled versions of one another. Differences in size exist but vary among the linear distances considered.

4.9.1 The EDMA-I hypothesis test

Details of the EDMA-I procedure were introduced by Lele and Richtsmeier (1991) and are based on the Union-Intersection principle (Casella and Berger, 1990). We begin by presenting an intuitive description of this testing procedure.

Assume that there is no biological variability, and that mean form matrices are available for the two populations under study. If the two mean forms are simply scaled versions of each other, then the form matrices representing the mean forms will differ by a 'scaling factor,' S. If this is the case, then the form difference matrix will consist of a single number, a constant, S. If the form difference matrix consists only of S and we divide the maximum entry in the form difference matrix by the minimum entry in it, the ratio will be equal to 1. Since we are dividing the maximum entry in the form difference matrix by the minimum entry, the resulting ratio can never be smaller than 1. Moreover, as this ratio deviates increasingly from the value of 1, the implication is that the forms under consideration are more and more different from one another.

In reality, it would be rare to have the form matrices of two forms differ by a constant, S. In practice, we do not have the true mean forms but only their estimates based on the observed landmark data. Even if the true mean forms are scaled versions of each other, their estimates may not be. In practice then, the ratio of the maximum entry of the form difference matrix to the minimum entry will most likely be different from 1. Our goal is to determine the probability of obtaining the observed (or a larger) maximum-to-minimum ratio value due to underlying biological variability when the two mean forms are, in fact, similar. This probability is known as the *p-value*. If the calculated *p-value* is small, we claim that the observed value is unlikely under the null hypothesis of similarity in form and reject the null hypothesis. By the same results, we entertain the alternative hypothesis. The EDMA-I test of the null hypothesis of similarity in forms assumes that the variances of the two samples being considered are equal.

The null hypothesis is that the average shapes of the two samples are the same. The steps for testing for the equality of average shapes are as follows. Again, let $A_1, A_2,..., A_n$ and $B_1, B_2,..., B_m$ be the landmark coordinate matrices for individuals from the two samples.

STEP 1. Estimate the mean form matrices for sample A and for sample B.

STEP 2. Calculate the form difference matrix for these samples, $FDM(B, A)$ and sort the vector in ascending order.

STEP 3. Calculate the test statistic, T, which is the ratio of the maximum ratio in the matrix $FDM(B, A)$ to the minimum entry in $FDM(B, A)$. Using our observations from samples A and B, $T_{obs} = \dfrac{\max_{ij} FDM_{ij}(B, A)}{\min_{ij} FDM_{ij}(B, A)}$.

The next step constitutes a Bootstrap approach to estimate the null distribution of the test statistic, T. To do this you must first choose one of the samples, say B, as your baseline sample.

STEP 4. Select n individuals randomly and with replacement from sample B, and call this sample A_1. Then, select m individuals randomly and with replacement from sample B, and call this sample B_1. Follow steps 1, 2, and 3 using A_1 and B_1 to obtain a T value for the bootstrapped sample.

STEP 5. Repeat Step 4 an adequate number of times (e.g., 200 to 1000 times).

STEP 6. Use the distribution of T values produced by Steps 4 and 5 in the form of a histogram as the null distribution of T when the null hypothesis is true. If T_{obs} falls in the upper $\alpha\%$ tail of the null distribution, we reject the null hypothesis at the $\alpha\%$ level of significance.

Notice that Step 4 requires the choice of a baseline sample. This choice is required because the test that we have designed does not simply test whether or not the two mean forms are similar. Instead, we test whether or not the average form of the first sample is similar to the average form of the baseline group. Our test is designed to answer the specific question, is the mean form of Group A similar to the mean form of Group B? Importantly, the answer to this question may not be the same as the answer to the question, is the mean form of Group B similar to the mean form of Group A? Our test is a one-way test and due to the design of the test, one group (generally the group with the larger sample size) must be chosen to serve as the baseline group.

The question that this test addresses is similar to the question, could sample A have arisen from the distribution of the population from which sample B was obtained? This subtle point that pertains to this testing procedure means that if one wants to know for certain if two populations are different from one another, the testing should be done twice using each of the samples as the baseline in one of the analyses. If one sample is particularly small, preventing its use as the baseline sample in testing, it is important to make this clear in reporting the results that the test is a one-way test. The EDMA-I test also requires the assumption that variances in the two samples are equal. If information is available that indicates that this assumption is not supported by the data, results of this test may be misleading.

4.9.2 A two-way test of the null hypothesis using EDMA-II

This procedure, discussed in detail by Lele and Cole (1996), is based on ideas outlined above, but uses a different statistic for testing the null hypothesis of similarity in shapes. An important difference between the two procedures is that unlike EDMA-I, EDMA-II does not require the assumption of equality of variances in the two populations. EDMA-II is also a two-way test removing the need to choose one of the samples as the baseline group. However, EDMA-II does require that a scaling factor be chosen to represent the "size" variable. A single linear distance, or a combination of all linear distances (e.g., the geometric mean of the distances) could be used to scale the data and represent size.

When testing the null hypothesis of similarity of shapes using the EDMA-II procedure, it is important to realize that choice of a measure to represent "size" has implications. Although the choice of the scaling factor does not affect the validity of the test, it does affect the power of the test (Lele and Cole, 1996). The logic for choosing a particular scaling factor should be biological and not based simply on mathematical convenience. Choice of the scaling factor will be influenced by the biological problem at hand. For example, the geometric mean might be a reasonable scaling factor for studying skulls, whereas overall length may be a more appropriate scaling factor in the study of femur morphology.

When conducting the EDMA-II test, the first step is to select a scaling factor. Using this scaling factor, a shape difference matrix is calculated. If the two mean forms are similar to each other, irrespective of which scaling factor is chosen, the shape difference matrix will consist of zeros. Thus, if the two forms are not similar, the entry that is farthest away from zero (either a positive or negative value) provides us with a measure of their dissimilarity.

The steps for testing for the equality of average shapes using

EDMA-II are as follows. Let $A_1, A_2,..., A_n$ and $B_1, B_2,..., B_m$ be the landmark coordinate matrices for individuals from the two samples.

> STEP 1. Estimate the mean form matrix for sample A and sample B.
>
> STEP 2. Calculate the scaling factor for each mean form matrix and use these scaling factors to standardize the mean form matrices for each sample by dividing each entry of the matrix by the respective scaling factor. For example, divide each entry of $FM(A)$ by the scaling factor estimated from sample A, S_A. The new matrix is called the shape matrix of A and is denoted as $SM(A)$. Do the same for sample B using S_B to produce $SM(B)$.
>
> STEP 3. Calculate the shape difference matrix for these samples, $SDM(B,A)$, by calculating the *difference* between like linear distances, and sort $SDM(B,A)$ written as a vector in ascending order. The test statistic Z is given by either the minimum entry in $SDM(B,A)$, or the maximum entry in $SDM(B,A)$, whichever is most different from zero.
>
> STEP 4. Using the estimates of the mean forms and variance-covariance matrices, generate parametric bootstrap (Monte Carlo) samples under the Gaussian perturbation model. Calculate the Z statistic for the bootstrap samples.
>
> STEP 5. Repeat Step 4 an adequate number of times (e.g., 200 times) to obtain $Z_1, Z_2, ..., Z_{200}$. Sort these values in ascending order and denote the ordered values by $Z_{(1)}, Z_{(2)}, ..., Z_{(200)}$. The $100 (1-\alpha)$% confidence interval is delimited by $(Z_{(11)}, Z_{(190)})$. If this interval contains 0, accept the null hypothesis that the samples are similar in shape; otherwise, the test provides evidence that the two forms are not similar.

The above testing procedure works well in most situations. However there is a situation where careless application of Step 5 can be problematic. Look carefully at the distribution of the values in SDM. If the maximum positive entry in SDM and the maximum negative entry in SDM have very similar magnitudes, it is necessary to look carefully at the histogram of $Z_1, Z_2, ..., Z_{200}$. If this histogram indicates a bimodal distribution with a large dip in the number of observations near zero, the null hypothesis should be rejected. Careless application and interpretation of Step 5 will lead to acceptance of the null hypothesis in this situation. As in all statistical analyses, the user must be wary of outcomes and pay close attention to data and distributions. The example analysis given later in this chapter demonstrates this type of situation and its resolution.

Table 4.1 summarizes the differences and similarities between the two methods for testing the hypothesis of similarity in shapes of forms sampled from two populations.

Table 4.1. Tabulation of the salient characteristics of the two methods for hypothesis testing using EDMA.

EDMA-I	EDMA-II
Description and comparison of forms are coordinate system-free	Description and comparison of forms are coordinate system-free
Null hypothesis states that the shapes are similar	Null hypothesis states that the shapes are similar
Enables statistical determination of differences due exclusively to scale or due to differences in form	Enables decomposition of statistically significant differences into differences in shape and differences in scaling factor (or "size")
Mean form matrices are invariant to the choice of scaling factor	Mean shape matrices are dependent on the choice of scaling factor
Assumes equality of the variance-covariance matrices for the two samples under consideration	Does not assume equality of the variance-covariance matrices
Testing procedure cannot determine localized differences; confidence intervals for elements of the FDM have been developed for this purpose.	Testing procedure cannot determine localized differences; confidence intervals for elements of the SDM have been developed for this purpose.

4.9.3. Cautionary notes regarding hypothesis testing
for similarity of shape

1. We emphasize that statistical testing for equality of shape, although a standard feature of most published analyses, is overrated and rarely provides the information desired or required by the scientist. This is because the answer given from a statistical test is limited to either, "yes, the shapes are the same" or "they are different." What is generally needed is the detection of those parts of the forms that are similar and those that are different, along with a measure of this difference. Localization facilitates the formulation of informed hypotheses about "why" or "how" two forms differ. Explanation and discovery are our goals, not description. Our method for the calculation of confidence intervals, in addition to the exploration procedures described below, are tools designed for studying the more subtle and meaningful aspects of form difference.

2. There are many different ways to test a hypothesis of equality of shape. Unfortunately, due to the nature of the problem, there cannot be a single test that is 'best' in every situation. By 'best' we mean the test that has the greatest statistical power to detect a real shape difference (the probability of rejecting a false null hypothesis). Simulation studies can be misleading because they report the power of one test relative to others in a *specific situation*. The problem is that one can always find a situation where any particular test behaves well. Due to the nature of biological data, simulation studies cannot support the use of one test over the other in all, or even most, situations. In the light of the above discussion we feel it necessary to remark on the misleading nature of the commentary on the EDMA testing procedures provided by James Rohlf (2000). As we have pointed out, superimposition and deformation approaches base their inference on non-identifiable parameters, which makes these approaches scientifically and statistically problematic, and possibly undesirable. Unfortunately, Rohlf (2000) provides a highly biased view of shape analysis in his paper. The main flaws in his argument are presented below.

    a) Using simulations, Rohlf shows that the EDMA testing

procedures can have smaller power as compared to some other testing procedures. As pointed out here, and in the original papers, neither EDMA-I or II are the uniformly most powerful (UMP) tests. We never claimed EDMA-I or EDMA-II to be UMP tests. Like most informed statisticians, we are aware of the fact that tests in multivariate analysis are seldom uniformly best; that is, they are not the most powerful alternative in every possible situation. Just as EDMA-based tests have smaller power in certain testing situations, Procrustes-based tests have smaller power in other situations. This finding alone does not make either testing procedure (or any alternative!) undesirable.

b)   Rohlf's claim that the EDMA-II procedure is invalid stems from an uninformed application of Step 5 in EDMA-II (see above). Rohlf has fallen prey to the error of blind statistical testing that was pointed out in the original paper by Lele and Cole (1996), and that we illustrate to the readers of this monograph by way of an example (see Section 4.1.2, Analysis of example data sets). EDMA-II, though not UMP, is currently the only test that can be applied when the covariance matrices for two populations are different.

c)   Rohlf chooses a particularly unrealistic covariance structure, namely isotropic covariance, to conduct his simulations. It has been shown that the Procrustes estimators of mean form, mean shape, and covariance are patently wrong for any covariance structure other than isotropic (Lele, 1993; Kent and Mardia, 1997). What is obvious to statisticians, but perhaps less clear to biologists, is that statistical inference procedures that use these erroneous estimators are not statistically valid. Rohlf (2000) ignores this important issue and chooses to concentrate on the non-UMP nature of the shape comparison tests.

4.10 Methods for exploring the form difference matrix

In the previous section, we showed how to test the null hypothesis of equality of shapes. Testing for the equality of shapes or forms is most often a formality; one almost required in published analyses, but one that may not provide information relevant to the problem under study. We have found in our experience that even when the null hypothesis of equality of overall shape or form cannot be rejected, there may be regions or loci where the forms are different. On the other hand, samples of forms that are shown to be statistically different may share features that are similar. What is needed is a method that localizes form difference to those parts of the objects that contribute significantly to the observed form difference, or that differentiate seemingly similar forms at very specific loci. This activity falls under the heading of *data exploration*. Calculating confidence intervals for elements of the form difference matrix is one way of uncovering those differences. In this section, we address the activity of *data exploration*. This includes methods for identifying those regions that contribute substantially to the observed form or shape difference and the ranking of parts of the object in terms of relative contribution to the observed difference.

When we specify the location of differences between forms under study, we say that the difference is "local" to a particular linear distance, to a particular landmark, or to a set of landmarks. We use the term "influential" to describe landmarks whose relative ranking is high in terms of the contribution to the difference in form or difference in shape. We remind the reader that the relative locations of landmarks are simply a manifestation of underlying biological processes. The landmarks are not "causing" the form difference or "influencing" the change in form. The difference occurs due to biological processes that affect the phenotype. Moreover, the difference noted local to a landmark might reflect differences local to a larger region that includes that particular landmark exclusively or several related landmarks. With this understanding, we will refer to "influential" landmarks, and designate differences that are "local" to particular regions.

4.10.1 Detection of influential landmarks

Consider the form difference matrix $FDM(B,A)$. Recall that if an entry in the form difference matrix is close to 1, it signifies that little difference exists local to the relative positions of the corresponding pair of landmarks. On the other hand, if the entry is markedly different from 1, it signifies that the relative position of the corresponding pair of landmarks is substantially different in the two mean forms representing the two populations. When an entry is very different from 1, this signifies that the two landmarks involved in defining this distance figure prominently in explaining the observed form difference.

The first step in exploring the form difference matrix is to arrange the entries in increasing order. Simply looking at the extremes (minimum and maximum) of such an ordered vector provides substantial information. If one or more distances at either the maximum or the minimum ends of the vector delineate a relevant region, this provides information relevant to the processes that underlie the form difference. However, analysis does not end with the simple reporting of which ratios are larger or smaller than 1. Instead, the researcher should attempt to relate these observations to potential underlying processes (i.e., physiological, biomechanical, pathological, evolutionary) and formulate hypotheses. The generation of new hypotheses from EDMA results will depend upon the scientific background of the investigator, the goals of the research, and the availability of alternate data. Once new hypotheses are generated, one can collect an alternate data set that bears on the new hypotheses. These data sets may or may not include landmark coordinates. The new hypotheses are then tested using the new data set. This process is similar to the general scientific process of hypothesis formulation followed by hypothesis testing, modification of the hypothesis on the basis of those results, additional data collection, and further hypothesis testing.

Two additional tools for systematic exploration of the form difference matrix are presented below. These tools can be used to gain an understanding of the observed form difference and as an aid in reformulation of hypotheses. These tools were originally presented in Lele and Richtsmeier (1992) and Cole and Richtsmeier (1998).

4.10.2 The landmark deletion approach.

The idea behind this technique is very simple. Recall the test statistic, T, defined in EDMA-I as $T = \dfrac{\max_{ij} FDM_{ij}(B,A)}{\min_{ij} FDM_{ij}(B,A)}$. Remember also that if the two mean forms are very similar to each other, the value of T is close to 1 and as the forms become more and more dissimilar, the value of T differs increasingly from 1.

Let us take a simple example. Consider the two objects shown in Figure 4.8 whose landmark coordinates are provided below.

$$\text{Let } A = \begin{bmatrix} 0 & 0 \\ 0 & 1 \\ 1 & 1 \\ 1 & 0 \end{bmatrix} \text{ and } B = \begin{bmatrix} 0 & 0 \\ -0.2 & 1.2 \\ 1.5 & 1.6 \\ 1.1 & 0 \end{bmatrix}.$$

The corresponding form matrices (with numbers rounded to two significant digits) are given by

$$FM(A) = \begin{bmatrix} 0 & 1 & 1.41 & 1 \\ 1 & 0 & 1 & 1.41 \\ 1.41 & 1 & 0 & 1 \\ 1 & 1.41 & 1 & 0 \end{bmatrix}$$

and

$$FM(B) = \begin{bmatrix} 0 & 1.22 & 2.19 & 1.1 \\ 1.22 & 0 & 1.75 & 1.77 \\ 2.19 & 1.75 & 0 & 1.65 \\ 1.1 & 1.77 & 1.65 & 0 \end{bmatrix}$$

The form difference matrix, in the vector form, can then be given as

$$FDM(A, B) = \begin{bmatrix} 1.22 \\ 1.55 \\ 1.10 \\ 1.75 \\ 1.25 \\ 1.65 \end{bmatrix}$$

The value of T for the FDM of these two objects is the ratio of the maximum entry (1.75) and the minimum entry (1.10) and hence $T=$ 1.59. Now suppose we delete landmark 3 from both objects and re-calculate the value of T for the objects that now consist only of landmarks 1, 2, and 4. The new value of T, 1.14, is much closer to 1 than the original T calculated using all landmarks. After deleting landmark 3, the value of T indicates that the remainders of the two objects are more

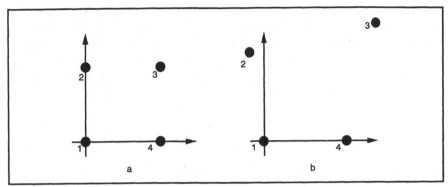

Figure 4.8. Graphical Representation of landmark location for objects A and B.

similar to each other than we found originally. We interpret this as a demonstration of the importance of landmark 3 in establishing the form difference between these two objects. Similarly if we calculate T after deleting landmark 4 (but include landmarks 1, 2, and 3), the value of T, is 1.44. This value of T is reduced as compared to the original T calculated using all landmarks, but the change in the value of T is not as large as when landmark 3 was deleted. From this experiment, we conclude that landmark 3 plays a larger part in determining the form difference than does landmark 4. Both landmarks are influential but landmark 3 is more influential than landmark 4.

If we begin this exercise again, but delete landmark 1, the value of T is 1.40. This value is similar to the one obtained after deleting landmark 4. This implies that landmarks 1 and 4 have comparable influence. The value of T obtained when deleting landmark 2 is 1.5, indicating that landmark 2 does not explain much of the difference between the two forms. From these analyses, we can rank the landmarks according to their influence in the differentiation of forms; landmark 3 being the most influential, landmarks 1 and 4 having similar influence and landmark 2 the least influential of all.

Based on this simple example, we can propose the following procedure for detecting influential landmarks.

STEP 1: Calculate the value of T for the complete object.

STEP 2: Delete the i-th landmark and recalculate the value of T. Let us denote it by $T_{(-1)}$. Calculate $T_{(-1)}, T_{(-2)}, \ldots, T_{(-K)}$.

One can consider a landmark to be the most influential if it substantially reduces the value of the original T upon deletion. The landmark that reduces the value of T to the next lowest value is the second-most influential landmark, and so on. Thus, one can rank the landmarks according to their influence in explaining the form difference. If testing for similarity in shapes is of concern, the p-value can be estimated and checked for each value of $T_{(-i)}$ noting those landmarks that make a substantial difference in the level of significance.

Instead of, or in addition to, working through this process one landmark at a time, groups of landmarks can be identified and deleted as a set. The object could be divided into regions or features, each defined by a set of landmarks, and the sets can be deleted one at a time. Parts of the object can then be ranked according to their role in influencing the form difference. Those parts of the object that reduce the value of T substantially when deleted are highly influential.

In summary, if a particular part of the object is responsible for a majority of the form difference, deletion of that part of the object should result in an improved match of the forms defined by the remaining landmarks. Using these observations, the influence of particular landmarks or sets of landmarks can be ranked by observing the reduction in T caused by deleting that particular set: the larger the reduction, the larger the influence.

Two concerns should be noted when applying this logic and this approach. First, remember that the *FDM* is based on the comparison of mean form matrices and that these are only estimates of the mean form, and not the true mean forms. The various values of T calculated during the landmark deletion process are not directly comparable due to sampling variability. We need a way to calibrate the T values so that sampling variability is taken into account. The p-values can be used towards this purpose. After deletion of a set of landmarks, a substantial *increase* in the p-value indicates substantial influence of the *deleted* landmarks. Recall that as T gets closer to 1, the corresponding p-value increases. Thus, any reduction in the value of T corresponds to an increase in the p-value.

Second, some consideration should be given to the number of landmarks deleted. To this end, common sense is our only guide. For example, if deletion of a single landmark or relatively small set of landmarks achieves a reduction in T (or increase in the p-value) similar to that achieved by the deletion of a larger set, the smaller set is more influential. That is, the smaller set accounts for a "statistically comparable difference," but because this difference is confined to a smaller area, the locus of the difference is more identifiable and hypotheses regarding processes responsible for the differences may be more specific.

We stress that these data exploration techniques should not be applied haphazardly. Statistical exploration for influential landmarks can be important scientifically, but biological knowledge, focused thinking, and common sense are necessary for understanding and explaining the differences observed, as well as guarding against spurious results. Investigators using these techniques should understand the underlying logic, and should always use their judgment when interpreting and explaining the implications of the results. Simplistic conclusions reporting localized increases and decreases in particular linear distances are rarely informative by themselves. Plausible processes that may underlie such changes should *always* be discussed. In the best case, the results of data exploration should be formulated as hypotheses to be tested using alternate, relevant data that may need to be generated through other experiments, or the analysis of additional specimens.

4.11 A graphical tool for the detection of influential landmarks

This method, proposed by Cole and Richtsmeier (1998), involves the use of a two-dimensional scatter plot to summarize, explore, and interpret a form difference matrix. Positions along the horizontal axis of this scatter plot correspond to the landmarks numbered from 1 to K. Positions along the vertical axis are real numbers corresponding to values of the off-diagonal elements of the form difference matrix for each landmark pair.

Let us continue with data presented earlier (Figure 4.8). Rewrite the form difference matrix for this comparison as a square, symmetric matrix. The first row of this matrix contains the ratios of the distances in the two objects that involve landmark 1 as one of the end points. Similarly, the second row of the matrix contains the ratios of the distances in two objects that involve landmark 2 as one of the end points.

Now consider the scatter plot in Figure 4.9. The landmark numbers 1, 2, 3, and 4 are plotted on the horizontal axis. Above the point corresponding to landmark 1, those values corresponding to the first row of the form difference matrix are plotted. Above the point corresponding to landmark 2, the ratios that correspond to the second row of the form difference matrix are plotted. The same is done for those ratios that correspond to the third and fourth row of the form difference matrix. Remember that the form difference matrix is a square symmetric matrix. This means that our scatter plot will contain two points for the

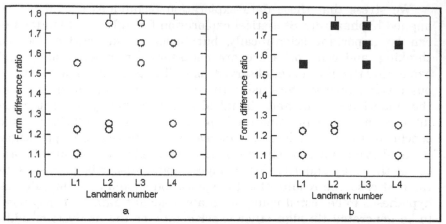

Figure 4.9. Scatter plot showing results of an EDMA comparison of the two forms whose data are presented in Figure 4.8. The Y-axis represents the value of the elements of the form difference matrix, while landmark numbers are shown along the X-axis. The ratio value for the comparison of each linear distance is shown twice on the graph, once for each landmark that is an endpoint for the linear distance. The values for the comparison of a linear distance with itself (that value always being "1") are not entered on the graph. On the graph shown at the left we see all form difference ratios for all linear distances. The column for landmark three suggests that large values are associated with this landmark. On the graph shown at the right all ratios associated with linear distances that have landmark 3 as an endpoint are graphed as closed squares showing that the outlying values for each landmark map to the distance between that landmark and landmark 3.

value corresponding to each landmark pair. For example, the value representing the ratio for the landmark pair 3, 4 will be plotted once above the "3" on the X-axis and again above the "4" on the X-axis.

A pattern is easily seen when looking at the scatter plot (Figure 4.9a). All points plotted in the column for landmark 3 are skewed towards the extremes of the values on the Y-axis. When a different marker (e.g., ■) is used to represent those points that correspond to linear distances that have landmark 3 as an endpoint (Figure 4.9b), these linear distances stand out from all others.

Cole and Richtsmeier (1998) suggest the use of such plots for exploring the influence of a particular landmark. The distribution of the form difference matrix elements that correspond to a single landmark gives a visual impression of its influence. This approach could be used as a complement to the "delete one" procedure just described, by sequentially eliminating landmarks from the plot beginning with the most influential. However, if this is done the investigator should be careful to preserve the horizontal and vertical scales of the scatter plot so that perceptions of influence are consistent from one plot to another (Cole and Richtsmeier, 1998).

Additional visualization tools suggested by Cole and Richtsmeier (1998) are best described using example data sets. In the following section, we explore the data sets introduced in chapter 1 using these tools.

4.12 Analysis of example data sets: mouse mandibles

In Chapter 1, we introduced the Ts65Dn mouse model for Down Syndrome. The three-dimensional coordinates of the normal and Ts65Dn mandibular data sets are presented in Chapter 3. Here, we use the methods outlined in the earlier part of this chapter to compare a sample of Ts65Dn mice mandibles (N=7) with a sample of normal littermates (N=13).

Form difference matrix

In this example, the mean form estimated from the normal mouse mandible is used as the numerator, and the estimated mean form of the Ts65Dn mandible is used as the denominator in an EDMA-I analysis. The null hypothesis is that the shapes of the two mean forms are similar. The form difference matrix is presented in matrix format (Table 4.2). Since the form difference matrix is symmetric, we present only the diagonal elements (all equal to zero) and lower off-diagonal elements.

Observations can be made by simple inspection of the form difference matrix. For example, it is easy to see that elements that are greater than one far out number elements that are less than one. This means that the normal mandible is generally larger than the Ts65Dn mandible along most dimensions. It also appears that many of the linear distances that have landmark 1 as an endpoint (look to the column labeled LND 1) are of relatively greater magnitude than elements associated with other landmarks.

Inspecting the form difference matrix in this way is not systematic and can become tedious, especially as the number of landmarks increases. One way to look for patterns among the linear distances in terms of their contribution to the difference in forms is to calculate confidence intervals for the linear distances.

Table 4.2. The form difference matrix for the comparison of the Ts65Dn and normal mouse mandibles

	LND 1	LND 2	LND 3	LND 4	LND 5	LND 6	LND 7	LND 8	LND 9	LND 10	LND 11
LANDMARK 1	0.000										
LANDMARK 2	1.090	0.000									
LANDMARK 3	0.793	1.147	0.000								
LANDMARK 4	0.883	1.150	1.021	0.000							
LANDMARK 5	1.147	1.102	1.029	1.032	0.000						
LANDMARK 6	1.141	1.044	1.017	1.015	0.987	0.000					
LANDMARK 7	1.133	1.051	1.024	1.024	1.011	1.055	0.000				
LANDMARK 8	1.150	1.067	1.036	1.034	1.030	1.068	1.054	0.000			
LANDMARK 9	1.170	1.070	1.025	1.025	1.014	1.108	1.020	1.038	0.000		
LANDMARK 10	1.196	1.068	1.018	1.019	0.994	1.053	1.040	1.076	1.000	0.000	
LANDMARK 11	1.259	1.084	1.011	1.014	1.006	1.054	1.040	1.060	1.050	1.031	0.000

90% bootstrap confidence intervals for form difference between Ts65Dn and normal mouse mandibles

Confidence intervals are calculated following methods outlined previously in this chapter and detailed in Part 2 of this chapter. Once confidence intervals are estimated, the form difference matrix is written as a vector, along with the lower and upper limits of a 90% confidence interval. Remember that if the confidence interval does not contain the value 1 (a value of 1 signifying that two forms are the same for a particular linear distance), then the null hypothesis that the two forms are similar for that linear distance is rejected. We have marked the linear distances that do not contain the value 1 within the confidence interval with a "*". Of the 55 linear distances, 39 of the confidence intervals do not contain the value 1.0. This allows us to pinpoint those measures that do not differ between the two samples as well as those that do. For example, many of the linear distances that have landmark 4 or 5 as endpoints do not differ between the samples.

Confidence intervals for form differences between Ts65Dn and normal mouse mandibles based on 100 resamples with α=0.10.

	Lower Limit	Estimate	Upper Limit
LANDMARK 2 with LANDMARK 1	1.063	1.090	1.111*
LANDMARK 3 with LANDMARK 1	0.767	0.793	0.815*
LANDMARK 3 with LANDMARK 2	1.108	1.147	1.184*
LANDMARK 4 with LANDMARK 1	0.870	0.884	0.894*
LANDMARK 4 with LANDMARK 2	1.093	1.150	1.210*
LANDMARK 4 with LANDMARK 3	0.982	1.021	1.050
LANDMARK 5 with LANDMARK 1	1.111	1.147	1.179*
LANDMARK 5 with LANDMARK 2	1.058	1.102	1.163*
LANDMARK 5 with LANDMARK 3	0.996	1.029	1.072
LANDMARK 5 with LANDMARK 4	1.004	1.032	1.067*
LANDMARK 6 with LANDMARK 1	1.108	1.142	1.163*
LANDMARK 6 with LANDMARK 2	1.027	1.044	1.061*
LANDMARK 6 with LANDMARK 3	1.002	1.018	1.033*
LANDMARK 6 with LANDMARK 4	0.995	1.015	1.034
LANDMARK 6 with LANDMARK 5	0.869	0.988	1.074
LANDMARK 7 with LANDMARK 1	1.099	1.133	1.156*
LANDMARK 7 with LANDMARK 2	1.031	1.051	1.067*
LANDMARK 7 with LANDMARK 3	1.008	1.025	1.038*
LANDMARK 7 with LANDMARK 4	0.999	1.024	1.040
LANDMARK 7 with LANDMARK 5	0.945	1.011	1.063

LANDMARK 7 with LANDMARK 6	1.019	1.056	1.093*
LANDMARK 8 with LANDMARK 1	1.112	1.151	1.182*
LANDMARK 8 with LANDMARK 2	1.042	1.067	1.093*
LANDMARK 8 with LANDMARK 3	1.015	1.036	1.057*
LANDMARK 8 with LANDMARK 4	1.012	1.035	1.056*
LANDMARK 8 with LANDMARK 5	0.968	1.031	1.077
LANDMARK 8 with LANDMARK 6	1.031	1.068	1.099*
LANDMARK 8 with LANDMARK 7	1.011	1.054	1.109*
LANDMARK 9 with LANDMARK 1	1.130	1.170	1.197*
LANDMARK 9 with LANDMARK 2	1.050	1.070	1.085*
LANDMARK 9 with LANDMARK 3	1.009	1.025	1.037*
LANDMARK 9 with LANDMARK 4	1.003	1.025	1.038*
LANDMARK 9 with LANDMARK 5	0.920	1.014	1.094
LANDMARK 9 with LANDMARK 6	1.069	1.108	1.141*
LANDMARK 9 with LANDMARK 7	0.980	1.020	1.068
LANDMARK 9 with LANDMARK 8	0.989	1.038	1.087
LANDMARK 10 with LANDMARK 1	1.163	1.196	1.217*
LANDMARK 10 with LANDMARK 2	1.049	1.068	1.080*
LANDMARK 10 with LANDMARK 3	1.007	1.018	1.026*
LANDMARK 10 with LANDMARK 4	1.002	1.019	1.028*
LANDMARK 10 with LANDMARK 5	0.918	0.995	1.053
LANDMARK 10 with LANDMARK 6	1.019	1.053	1.079*
LANDMARK 10 with LANDMARK 7	1.001	1.041	1.078*
LANDMARK 10 with LANDMARK 8	1.025	1.077	1.132*
LANDMARK 10 with LANDMARK 9	0.941	1.000	1.062
LANDMARK 11 with LANDMARK 1	1.190	1.259	1.313*
LANDMARK 11 with LANDMARK 2	1.059	1.084	1.104*
LANDMARK 11 with LANDMARK 3	0.992	1.012	1.030
LANDMARK 11 with LANDMARK 4	0.994	1.015	1.031
LANDMARK 11 with LANDMARK 5	0.927	1.006	1.057
LANDMARK 11 with LANDMARK 6	1.021	1.054	1.079*
LANDMARK 11 with LANDMARK 7	1.007	1.041	1.068*
LANDMARK 11 with LANDMARK 8	1.016	1.060	1.099*
LANDMARK 11 with LANDMARK 9	1.012	1.050	1.089*
LANDMARK 11 with LANDMARK 10	0.967	1.031	1.109

Observe the first entry in the vector written above. The estimate 1.090 is greater than 1, indicating that the linear distance from landmark 1 to 2 is larger in the mean normal mandible (numerator) than in the mean Ts65Dn mandible (denominator). Notice that the confidence interval for this linear distance does not contain the value 1. A confidence interval that does not contain the value 1 indicates that the two forms are different for that specific linear distance. Confidence intervals for distances between landmarks 3 and 4, and 3 and 5 contain the value 1 and indicate that these distances are not significantly different in the two samples using the $\alpha=0.10$ confidence limits.

Next, consider the confidence interval calculated for the distance between landmarks 1 and 3. This interval does not include 1, but unlike the previous estimate examined, this estimate is less than 1. We interpret this as indicating that the distance between landmarks 1 and 3 is significantly smaller in the mean normal hemi-mandible relative to the mean Ts65Dn hemi-mandible.

Testing the null hypothesis of similarity in form using EDMA-I

One way to organize EDMA output is to write the form difference matrix as a vector and sort the elements from smallest to largest values. This format is shown below. It is convenient to write the output in this format when the EDMA-I test is used to determine whether or not we can reject the null hypothesis of similarity in shape. The Bootstrap test, described previously in this chapter, is applied and the output is given below.

Form Difference Matrix (Sorted from Minimum to Maximum)

LANDMARK 3 with LANDMARK 1	0.79328
LANDMARK 4 with LANDMARK 1	0.88377
LANDMARK 6 with LANDMARK 5	0.98774
LANDMARK 10 with LANDMARK 5	0.99459
LANDMARK 10 with LANDMARK 9	1.00024
LANDMARK 11 with LANDMARK 5	1.00576
LANDMARK 7 with LANDMARK 5	1.01100
LANDMARK 11 with LANDMARK 3	1.01177
LANDMARK 9 with LANDMARK 5	1.01416
LANDMARK 11 with LANDMARK 4	1.01458
LANDMARK 6 with LANDMARK 4	1.01521
LANDMARK 6 with LANDMARK 3	1.01765
LANDMARK 10 with LANDMARK 3	1.01787
LANDMARK 10 with LANDMARK 4	1.01864
LANDMARK 9 with LANDMARK 7	1.02009
LANDMARK 4 with LANDMARK 3	1.02135
LANDMARK 7 with LANDMARK 4	1.02390
LANDMARK 7 with LANDMARK 3	1.02461
LANDMARK 9 with LANDMARK 3	1.02535
LANDMARK 9 with LANDMARK 4	1.02547
LANDMARK 5 with LANDMARK 3	1.02934
LANDMARK 8 with LANDMARK 5	1.03074
LANDMARK 11 with LANDMARK 10	1.03110
LANDMARK 5 with LANDMARK 4	1.03244
LANDMARK 8 with LANDMARK 4	1.03452

LANDMARK 8 with LANDMARK 3	1.03625
LANDMARK 9 with LANDMARK 8	1.03834
LANDMARK 10 with LANDMARK 7	1.04054
LANDMARK 11 with LANDMARK 7	1.04073
LANDMARK 6 with LANDMARK 2	1.04363
LANDMARK 11 with LANDMARK 9	1.05043
LANDMARK 7 with LANDMARK 2	1.05127
LANDMARK 10 with LANDMARK 6	1.05319
LANDMARK 8 with LANDMARK 7	1.05413
LANDMARK 11 with LANDMARK 6	1.05423
LANDMARK 7 with LANDMARK 6	1.05551
LANDMARK 11 with LANDMARK 8	1.06014
LANDMARK 8 with LANDMARK 2	1.06728
LANDMARK 8 with LANDMARK 6	1.06791
LANDMARK 10 with LANDMARK 2	1.06810
LANDMARK 9 with LANDMARK 2	1.06964
LANDMARK 10 with LANDMARK 8	1.07662
LANDMARK 11 with LANDMARK 2	1.08393
LANDMARK 2 with LANDMARK 1	1.09016
LANDMARK 5 with LANDMARK 2	1.10249
LANDMARK 9 with LANDMARK 6	1.10790
LANDMARK 7 with LANDMARK 1	1.13338
LANDMARK 6 with LANDMARK 1	1.14152
LANDMARK 5 with LANDMARK 1	1.14671
LANDMARK 3 with LANDMARK 2	1.14706
LANDMARK 4 with LANDMARK 2	1.15007
LANDMARK 8 with LANDMARK 1	1.15058
LANDMARK 9 with LANDMARK 1	1.17023
LANDMARK 10 with LANDMARK 1	1.19642
LANDMARK 11 with LANDMARK 1	1.25924

T_{obs} [=max/min of FDM or 1.25/0.79] = 1.58
p-value= 0.005

Since T_{obs}, the observed value of T calculated from the data, lies in the extreme tail of the distribution of bootstrapped T-values, we reject the null hypothesis of similarity in shape. Remember that the p-value represents the probability of obtaining the observed value of T when the two populations are identical in shape. Remember also that this is a one-way test. We selected the normal sample as the baseline group because of its larger sample size. This test enables us to formally state that the average form of the Ts65Dn mandible (test sample) is significantly different from the average form of the normal sample (baseline group). If sample size permits, and the investigator believes it necessary, the test can be run again using the alternate sample as the baseline sample.

Table 4.3. Distribution of Bootstrapped T Statistics:

T	
1.028	|*****
1.057	|***
1.087	|***
1.116	|***
1.146	|***********************
1.175	|************
1.205	|***
1.234	|****
1.264	|
1.293	|
1.322	|
1.352	|
1.381	|
1.411	|
1.440	|
1.470	|
1.499	|
1.528	|
1.558	|
1.587	|T_{obs}

Testing the null hypothesis of similarity in form using EDMA-II

EDMA-II was developed for testing the difference in shapes of two populations where the variance-covariance matrices are not equal. As described in Chapter 1, the Ts65Dn mouse is a model for Down syndrome. Previous quantitative studies of the craniofacial phenotype in children with Down syndrome have documented increased variability in this group as compared to unaffected children (Kisling 1966; Thelander and Pryor, 1966; Frostad, Cleall et al., 1971; Cronk and Reed, 1981). Because Ts65Dn mice are at dosage imbalance for many of the same genes duplicated in Down syndrome (Richtsmeier, Baxter et al., 2000), it may be necessary to adopt a working hypothesis that the Ts65Dn mouse will also show increased variability as compared to euploid littermates. If evidence for the working hypothesis is found, application of EDMA-I to these samples may be inappropriate and EDMA-II testing should be implemented.

Using the estimators of mean form and variance-covariance structure described above, we estimate the mean shapes for the two populations and calculate the *shape difference*. In the example below, we use the geometric mean of the distances among mandibular landmarks as the scaling factor and standardize the form difference matrix for a sample by dividing each entry by the scaling factor calculated for that sample. The result is a mean shape matrix *that is dependent upon the choice* of scaling factor. In other words, if a scientific reason exists for choosing an alternate measure as a scaling factor (i.e., the length of a particular linear distance), the mean shape matrix will change. The null hypothesis that the shapes are similar is tested using a parametric Bootstrap (i.e., Monte Carlo) procedure (see previous description of EDMA-II this section and Part 2 of this chapter).

This example uses the same landmark data from the Ts65Dn and normal mandibles used in the EDMA-I example above. The mean form matrices estimated for these samples were given in Chapter 2. The scaling variables for the two samples are:

Geometric mean for linear distances from the normal sample: 5.23902
Geometric mean for linear distances from the Ts65Dn sample: 4.98840
Difference in geometric means between samples: 0.25062

Dividing each entry of the mean form matrix estimated for each sample by the sample-specific scaling factors, we arrive at a shape matrix for each sample.

Elements of the shape difference matrix represent the difference between like-linear distances in the two samples:

Shape matrix for the normal mandibles (each linear distance divided by the geometric mean):

	LND 1	LND 2	LND 3	LND 4	LND 5	LND 6	LND 7	LND 8	LND 9	LND 10	LND 11
LND 1	0.00000										
LND 2	1.28673	0.00000									
LND 3	0.40283	0.96729	0.00000								
LND 4	0.75299	0.80826	0.35283	0.00000							
LND 5	1.07724	1.06845	1.10641	1.29204	0.00000						
LND 6	1.39685	1.41786	1.47406	1.67151	0.41324	0.00000					
LND 7	1.78392	2.12376	1.98717	2.25082	1.06821	0.72240	0.00000				
LND 8	1.75649	2.35898	2.03003	2.33225	1.31861	1.05632	0.46110	0.00000			
LND 9	1.34111	1.74983	1.54388	1.81640	0.69138	0.47966	0.45293	0.63439	0.00000		
LND 10	1.13230	1.72623	1.37657	1.67467	0.72994	0.64785	0.66296	0.67199	0.26851	0.00000	
LND 11	0.82424	1.44277	1.04616	1.34444	0.55934	0.67873	0.96170	1.00431	0.51771	0.33480	0.00000

Shape matrix for the Ts65Dn mandibles:

	LND 1	LND 2	LND 3	LND 4	LND 5	LND 6	LND 7	LND 8	LND 9	LND 10	LND 11
LND 1	0.00000										
LND 2	1.23961	0.00000									
LND 3	0.53332	0.88564	0.00000								
LND 4	0.89483	0.73810	0.36281	0.00000							
LND 5	0.98661	1.01781	1.12888	1.31431	0.00000						
LND 6	1.28516	1.42684	1.52127	1.72919	0.43939	0.00000					
LND 7	1.65306	2.12167	2.03688	2.30872	1.10967	0.71879	0.00000				
LND 8	1.60331	2.32131	2.05743	2.36769	1.34356	1.03884	0.45939	0.00000			
LND 9	1.20360	1.71808	1.58135	1.86028	0.71598	0.45470	0.46631	0.64166	0.00000		
LND 10	0.99395	1.69737	1.42035	1.72662	0.77078	0.64604	0.66914	0.65553	0.28193	0.00000	
LND 11	0.68744	1.39793	1.08593	1.39170	0.58408	0.67616	0.97048	0.99493	0.51762	0.34101	0.00000

Shape difference matrix:

LND 1	0.00000									
LND 2	0.04712	0.00000								
LND 3	-0.13049	0.08164	0.00000							
LND 4	-0.14184	0.07016	-0.00998	0.00000						
LND 5	0.09063	0.05064	-0.02247	-0.02227	0.00000					
LND 6	0.11170	-0.00898	-0.04721	-0.05767	-0.02615	0.00000				
LND 7	0.13085	0.00209	-0.04971	-0.05790	-0.04146	0.00361	0.00000			
LND 8	0.15318	0.03767	-0.02740	-0.03544	-0.02495	0.01748	0.00170	0.00000		
LND 9	0.13751	0.03174	-0.03747	-0.04388	-0.02460	0.02496	-0.01339	-0.00727	0.00000	
LND 10	0.13835	0.02886	-0.04378	-0.05195	-0.04084	0.00181	-0.00618	0.01647	-0.01342	0.00000
LND 11	0.13680	0.04484	-0.03978	-0.04726	-0.02474	0.00257	-0.00879	0.00938	0.00009	-0.00622	0.00000

Shape-Difference Matrix Written in Vector Format Sorted
From Minimum to Maximum:

LND 1	LND 4	-0.14184
LND 1	LND 3	-0.13049
LND 4	LND 7	-0.05790
LND 4	LND 6	-0.05767
LND 4	LND 10	-0.05195
LND 3	LND 7	-0.04971
LND 4	LND 11	-0.04726
LND 3	LND 6	-0.04721
LND 4	LND 9	-0.04388
LND 3	LND 10	-0.04378
LND 5	LND 7	-0.04146
LND 5	LND 10	-0.04084
LND 3	LND 11	-0.03978
LND 3	LND 9	-0.03747
LND 4	LND 8	-0.03544
LND 3	LND 8	-0.02740
LND 5	LND 6	-0.02615
LND 5	LND 8	-0.02495
LND 5	LND 11	-0.02474
LND 5	LND 9	-0.02460
LND 3	LND 5	-0.02247
LND 4	LND 5	-0.02227
LND 9	LND 10	-0.01342
LND 7	LND 9	-0.01339
LND 3	LND 4	-0.00998
LND 2	LND 6	-0.00898
LND 7	LND 11	-0.00879
LND 8	LND 9	-0.00727
LND 10	LND 11	-0.00622
LND 7	LND 10	-0.00618
LND 9	LND 11	0.00009
LND 7	LND 8	0.00170
LND 6	LND 10	0.00181
LND 2	LND 7	0.00209
LND 6	LND 11	0.00257
LND 6	LND 7	0.00361
LND 8	LND 11	0.00938
LND 8	LND 10	0.01647
LND 6	LND 8	0.01748
LND 6	LND 9	0.02496
LND 2	LND 10	0.02886
LND 2	LND 9	0.03174
LND 2	LND 8	0.03767
LND 2	LND 11	0.04484
LND 1	LND 2	0.04712

LND 2	LND 5	0.05064
LND 2	LND 4	0.07016
LND 2	LND 3	0.08164
LND 1	LND 5	0.09063
LND 1	LND 6	0.11170
LND 1	LND 7	0.13085
LND 1	LND 11	0.13680
LND 1	LND 9	0.13751
LND 1	LND 10	0.13835
LND 1	LND 8	0.15318

Z statistic: 0.15318

In this example, the element at the positive end of the vector is farther from 1 than the element at the negative end of the vector (the shape difference between LND 1 and LND 8 = 0.15318 while the shape difference between LND 1 and LND 4 = $-.14184$), and so the former is chosen as the Z statistic. Using the parametric bootstrap procedure described previously, a 90% confidence interval for the Z statistic is estimated and shown below.

Table 4.4. Confidence Interval for Z with $\alpha = 0.100$

Distribution of Bootstrapped Z Statistics	
-0.179	| *
-0.158	|***********
-0.138	|*********
-0.118	|
-0.098	|
-0.078	|
-0.058	|
-0.037	|
-0.017	|
0.003	|
0.023	|
0.043	|
0.064	|
0.084	|
0.104	|
0.124	|****
0.144	|Z*****************************
0.164	|*******************************
0.185	|************
0.205	| *

Upper bound: 0.18138   Lower bound: -0.16500

Since this confidence interval contains zero, these results do not allow us to reject the null hypothesis of similarity in shapes between the two samples. This result is in contrast to the result obtained using EDMA-I. Can this be explained? Recall that at the end of the description of EDMA-II, we remarked that if the magnitudes of the extreme negative element and the extreme positive element are similar, careful attention must be paid while applying and interpreting the results of EDMA-II. Observe that the bootstrap distribution of the test statistics is bimodal with a dip near zero. This suggests that the two shapes are different from each other, but because the two ends of the shape difference matrix are so similar, the test cannot determine if it is the positive end or the negative end that is more important. Instead of providing a summary measure that conveys this information, the summary measure tells the investigator that the forms are not different in shape, when they are clearly different. When the distribution of bootstrapped Z statistics is bimodal, the magnitudes of the extreme negative element and the extreme positive element should be checked for similarity in absolute value, and additional data exploration techniques should be used. This is an additional reason why we stress the use of the confidence interval procedure and advise against the isolated testing of a simple null hypothesis of equality of shapes.

In a case like this one where evidence points to a difference in the form of the two populations that are being compared, we may want to determine the role of scale in the difference. To examine this aspect of difference in form, confidence intervals for difference in scale (size) of the two populations can be estimated. To do this, we follow the same methods for generating bootstrapped samples as described above. For each pair of bootstrapped samples, the absolute difference in scaling variable is calculated. A distribution of the scaling variables is obtained in this way. A confidence interval can be obtained by sorting these bootstrapped scaling variables in ascending order. If this confidence interval contains zero, it indicates that there is no significant difference in the size of the forms representing the two populations (according to the chosen measure of size). For the example above, we calculate the confidence interval for the difference in the geometric means.

Table 4.5. Confidence Interval for Scale Difference with $\alpha = 0.100$

Distribution of Bootstrapped Scale Differences	
-1.961	|**
0.146	|**
0.160	|**
0.174	|**
0.187	|***
0.201	|*****
0.214	|*********
0.228	|*************
0.242	|***************
0.255	|D********
0.269	|**************
0.283	|*********
0.296	|*****
0.310	|*****
0.323	|**
0.337	|**
0.351	|*
0.364	|
0.378	|*
0.391	|
0.405	|*

Upper bound: 0.32669   Lower bound: 0.15548

This confidence interval (Table 4.5) does not contains zero and therefore suggests that the two populations do not differ significantly in scale. Remember, however, that there is no single value that represents "size" and that this result may change depending upon the chosen scaling factor. With no evidence for a difference in size, confidence intervals for the estimated shape difference matrix can be examined.

Confidence Intervals for Shape Differences Based on 100 Resamples with $\alpha = 0.10$

		Lower Limit	Estimate	Higher Limit
LND 1	LND 2	0.003	0.047	0.076
LND 1	LND 3	-0.159	-0.130	-0.112
LND 2	LND 3	0.055	0.082	0.107
LND 1	LND 4	-0.165	-0.142	-0.123
LND 2	LND 4	0.042	0.070	0.092
LND 3	LND 4	-0.027	-0.010	0.004

LND 1	LND 5	0.054	0.091	0.127
LND 2	LND 5	-0.003	0.051	0.085
LND 3	LND 5	-0.068	-0.022	0.021
LND 4	LND 5	-0.076	-0.022	0.019
LND 1	LND 6	0.072	0.112	0.136
LND 2	LND 6	-0.028	-0.009	0.009
LND 3	LND 6	-0.080	-0.047	-0.023
LND 4	LND 6	-0.093	-0.058	-0.034
LND 5	LND 6	-0.064	-0.026	0.014
LND 1	LND 7	0.099	0.131	0.157
LND 2	LND 7	-0.016	0.002	0.021
LND 3	LND 7	-0.076	-0.050	-0.032
LND 4	LND 7	-0.085	-0.058	-0.036
LND 5	LND 7	-0.084	-0.041	0.005
LND 6	LND 7	-0.014	0.004	0.019
LND 1	LND 8	0.126	0.153	0.181
LND 2	LND 8	0.017	0.038	0.068
LND 3	LND 8	-0.044	-0.027	-0.009
LND 4	LND 8	-0.057	-0.035	-0.018
LND 5	LND 8	-0.075	-0.025	0.026
LND 6	LND 8	-0.010	0.017	0.044
LND 7	LND 8	-0.014	0.002	0.016
LND 1	LND 9	0.106	0.138	0.159
LND 2	LND 9	0.014	0.032	0.053
LND 3	LND 9	-0.059	-0.037	-0.020
LND 4	LND 9	-0.064	-0.044	-0.023
LND 5	LND 9	-0.062	-0.025	0.015
LND 6	LND 9	0.013	0.025	0.038
LND 7	LND 9	-0.026	-0.013	-0.005
LND 8	LND 9	-0.024	-0.007	0.010
LND 1	LND 10	0.117	0.138	0.158
LND 2	LND 10	0.005	0.029	0.053
LND 3	LND 10	-0.064	-0.044	-0.026
LND 4	LND 10	-0.072	-0.052	-0.032
LND 5	LND 10	-0.086	-0.041	-0.003
LND 6	LND 10	-0.017	0.002	0.016
LND 7	LND 10	-0.023	-0.006	0.007
LND 8	LND 10	-0.005	0.016	0.033
LND 9	LND 10	-0.026	-0.013	-0.002
LND 1	LND 11	0.107	0.137	0.159
LND 2	LND 11	0.022	0.045	0.065
LND 3	LND 11	-0.062	-0.040	-0.019
LND 4	LND 11	-0.071	-0.047	-0.023
LND 5	LND 11	-0.064	-0.025	0.011
LND 6	LND 11	-0.016	0.003	0.021
LND 7	LND 11	-0.026	-0.009	0.006
LND 8	LND 11	-0.018	0.009	0.031
LND 9	LND 11	-0.016	0.000	0.013
LND 10	LND 11	-0.024	-0.006	0.011

Most confidence intervals excepting those that have landmarks 1 or 2 as an end point include zero within the confidence interval. This result localizes differences in shape to two landmarks, the coronoid process (landmark 1) and the mandibular angle (landmark 2).

A non-statistical approach to exploring the shape or form difference

Simple inspection of the form difference matrix written as a sorted vector (see section for testing the null hypothesis using EDMA-I above) makes certain observations evident. First, the two smallest values within the form different matrix are considerably smaller than any other values in the matrix. Second, these two values are associated with linear distances that share landmark 1 as an endpoint. Third, the normal mouse (the numerator sample) is equal to or larger than the Ts65Dn mouse for all other linear distances. This informs us that the distance between landmarks 1 and 3 and landmarks 1 and 4 are markedly different between the two samples, but also differ from the overall pattern of form difference between the samples.

Look now to the other extreme of the vector. Landmark 1 also appears as an endpoint for many of the linear distances that have relatively large magnitudes. Since the normal mouse is being used as the numerator in this calculation, linear distances at this end of the matrix are relatively larger in the mandibles of the normal mice. Clearly, landmark 1 is being revealed as a landmark that is influential to the difference between the mandibular morphology of these two populations. Our goal is to explain how some linear distances that have landmark 1 as an endpoint are smaller than normal in the Ts65Dn mandible while others are larger than normal.

Pictures of the mean forms can be helpful in interpreting this output. Consider Figure 4.10. In this figure we indicate those linear distances associated with landmark 1 that are larger in the Ts65Dn aneuploid mandible (from 1 to 3 and from 1 to 4) using a bold line. Those linear distances associated with landmark 1 that are smaller in the Ts65Dn aneuploid mandible are shown using a thin line. Landmark 1 is located on the apex of a process that is a site of attachment for one of the larger muscles of mastication. Linear distances originating from the coronoid process (landmark 1) and stretching along a posteroinferior axis are larger, while linear distances originating at landmark 1 but stretching along an anteroinferior axis are smaller in the Ts65Dn mandible. We propose that a local change in the

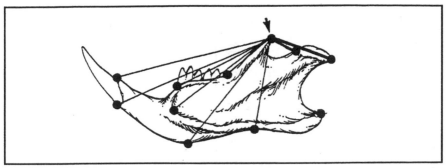

Figure 4.10. Differences in mandibular morphology between Ts65Dn aneuploid mice and their normal littermates. Thin lines indicate those linear distances that are smaller in the aneuploid individuals, while the bold lines indicate those linear distances that are larger in the aneuploid individuals. Interpretation of these results are that the morphology of the coronoid process (landmark 1, indicated by arrow) has changed such that the tip has moved more anteriorly and inferiorly.

morphology of the coronoid process has occurred that results in reduction of the overall process and anterior migration of its apex. Such a local change in morphology can account for the specific pattern of differences in linear distances on the mouse mandible.

The delete-one landmark approach

Another way to determine the influence of a particular landmark on the overall form difference is to re-run the EDMA-I or EDMA-II analysis after deleting one or more landmarks. Since we have found landmark 1 to be at both the minimum and maximum extremes of the form difference matrix calculated using all eleven landmarks, we run the EDMA-I analysis again, after deleting landmark 1.

Inspection of the form difference matrix indicates that the number of linear distances that are less than 1 has decreased. Below, we write the form difference matrix as a vector and observe the minimum and maximum values.

Form Difference matrix with landmark 1 deleted written as a sorted vector:

LND 6	LND 5	0.98568
LND 10	LND 5	0.99466
LND 10	LND 9	1.00052
LND 11	LND 5	1.00662

Form difference matrix for the normal to Ts65Dn mandible comparison with landmark 1 deleted:

	LND 2	LND 3	LND 4	LND 5	LND 6	LND 7	LND 8	LND 9	LND 10	LND 11
LND 2	0.00000									
LND 3	1.14628	0.00000								
LND 4	1.15038	1.01674	0.00000							
LND 5	1.10241	1.02920	1.03239	0.00000						
LND 6	1.04358	1.01710	1.01522	0.98568	0.00000					
LND 7	1.05127	1.02451	1.02395	1.01102	1.05524	0.00000				
LND 8	1.06729	1.03612	1.03454	1.03070	1.06781	1.05422	0.00000			
LND 9	1.06964	1.02529	1.02543	1.01432	1.10527	1.02059	1.03828	0.00000		
LND 10	1.06811	1.01778	1.01859	0.99466	1.05230	1.04098	1.07670	1.00052	0.00000	
LND 11	1.08413	1.01178	1.01459	1.00662	1.05392	1.04132	1.06053	1.05073	1.03194	0.00000

LND 7	LND 5	1.01102
LND 11	LND 3	1.01178
LND 9	LND 5	1.01432
LND 11	LND 4	1.01459
LND 6	LND 4	1.01522
LND 4	LND 3	1.01674
LND 6	LND 3	1.01710
LND 10	LND 3	1.01778
LND 10	LND 4	1.01859
LND 9	LND 7	1.02059
LND 7	LND 4	1.02395
LND 7	LND 3	1.02451
LND 9	LND 3	1.02529
LND 9	LND 4	1.02543
LND 5	LND 3	1.02920
LND 8	LND 5	1.03070
LND 11	LND 10	1.03194
LND 5	LND 4	1.03239
LND 8	LND 4	1.03454
LND 8	LND 3	1.03612
LND 9	LND 8	1.03828
LND 10	LND 7	1.04098
LND 11	LND 7	1.04132
LND 6	LND 2	1.04358
LND 11	LND 9	1.05073
LND 7	LND 2	1.05127
LND 10	LND 6	1.05230
LND 11	LND 6	1.05392
LND 8	LND 7	1.05422
LND 7	LND 6	1.05524
LND 11	LND 8	1.06053
LND 8	LND 2	1.06729
LND 8	LND 6	1.06781
LND 10	LND 2	1.06811
LND 9	LND 2	1.06964
LND 10	LND 8	1.07670
LND 11	LND 2	1.08413
LND 5	LND 2	1.10241
LND 9	LND 6	1.10527
LND 3	LND 2	1.14628
LND 4	LND 2	1.15038

T statistic [=max/min]: 1.16709

Notice that the value of the observed T statistic, T_{obs}, has been reduced from 1.587 (when all landmarks were included in the analysis) to 1.167. This indicates that landmark 1 was important to the overall form difference. The test of the null hypothesis using the reduced landmark set indicates the same finding.

p-value = 0.06

Table 4.7. Distribution of Bootstrapped T Statistics

1.034	|****
1.044	|***********
1.054	|***************
1.064	|**********************
1.074	|*********************
1.085	|****************************
1.095	|********************
1.105	|*****************
1.115	|***********
1.125	|******************
1.136	|*******
1.146	|*****
1.156	|**********
1.166	|**T_{obs}
1.176	|****
1.186	|***
1.197	|**
1.207	|
1.217	|*
1.227	|*

When confidence intervals are estimated for the reduced set of landmarks (see below), the exact values of the confidence intervals will not be the same as when these are estimated with all eleven landmarks. This occurs due to changes in the estimates for $\Sigma_K{}^*$ or the boostrapping procedures when landmarks are added to or subtracted from analysis (review Chapter 3).

90% confidence intervals for form differences between Ts65Dn and normal mouse mandibles after deleting landmark1:

	Low	Estimate	High
LND 3 LND 2	1.105	1.146	1.179
LND 4 LND 2	1.091	1.150	1.215
LND 4 LND 3	0.984	1.017	1.056
LND 5 LND 2	1.053	1.102	1.160
LND 5 LND 3	1.005	1.029	1.063
LND 5 LND 4	1.007	1.032	1.068

LND 6 LND 2	1.027	1.044	1.058
LND 6 LND 3	1.001	1.017	1.029
LND 6 LND 4	0.998	1.015	1.028
LND 6 LND 5	0.881	0.986	1.081
LND 7 LND 2	1.035	1.051	1.065
LND 7 LND 3	1.011	1.025	1.038
LND 7 LND 4	1.008	1.024	1.040
LND 7 LND 5	0.939	1.011	1.068
LND 7 LND 6	1.017	1.055	1.092
LND 8 LND 2	1.046	1.067	1.087
LND 8 LND 3	1.017	1.036	1.056
LND 8 LND 4	1.013	1.035	1.054
LND 8 LND 5	0.976	1.031	1.080
LND 8 LND 6	1.038	1.068	1.101
LND 8 LND 7	1.015	1.054	1.112
LND 9 LND 2	1.052	1.070	1.084
LND 9 LND 3	1.009	1.025	1.038
LND 9 LND 4	1.010	1.025	1.041
LND 9 LND 5	0.922	1.014	1.099
LND 9 LND 6	1.057	1.105	1.145
LND 9 LND 7	0.993	1.021	1.057
LND 9 LND 8	1.002	1.038	1.090
LND 10 LND 2	1.051	1.068	1.082
LND 10 LND 3	1.006	1.018	1.028
LND 10 LND 4	1.006	1.019	1.030
LND 10 LND 5	0.919	0.995	1.059
LND 10 LND 6	1.021	1.052	1.084
LND 10 LND 7	1.006	1.041	1.074
LND 10 LND 8	1.032	1.077	1.131
LND 10 LND 9	0.940	1.001	1.055
LND 11 LND 2	1.063	1.084	1.106
LND 11 LND 3	0.989	1.012	1.031
LND 11 LND 4	0.994	1.015	1.033
LND 11 LND 5	0.932	1.007	1.068
LND 11 LND 6	1.026	1.054	1.077
LND 11 LND 7	1.013	1.041	1.065
LND 11 LND 8	1.029	1.061	1.094
LND 11 LND 9	1.011	1.051	1.086
LND 11 LND 10	0.967	1.032	1.105

A simple graphical technique

Another way to determine the influence of particular landmarks is to view the elements of the form difference matrix as a scatter plot. This approach provides a visual impression of which landmarks are most influential in determining the difference between forms. Figure 4.11

provides such a scatter plot for the comparison of the Ts65Dn and normal mouse mandibles using all eleven landmarks. Landmark numbers are written along the horizontal axis while values of the off-diagonal elements of the form difference matrix (the form difference ratios) are on the vertical axis. Remember that since the form difference matrix is a square, symmetric matrix, the ratio for each linear distance appears twice. The value of any particular ratio will appear once in the column for each landmark that marks an endpoint of any linear distance.

In our example, the greatest dispersion of matrix values associated with a single landmark is seen local to landmark 1. If we substitute an alternate symbol for ratios corresponding to linear distances that have landmark 1 as an endpoint, the influence of landmark 1 to the form difference becomes even more obvious (Figure 4.11b). The largest matrix values associated with landmarks 4 through 11 involve landmark 1, and the smallest matrix values associated with landmarks 3 and 4 involve landmark 1. Visual inspection of this scatter plot provides the researcher with evidence to pursue the role of landmark 1 in determining the difference between these two populations of forms.

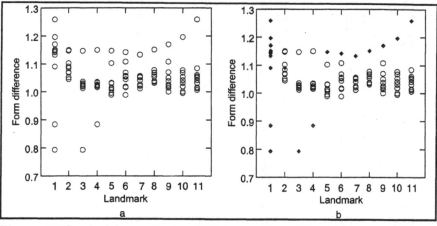

Figure 4.11 Graph of the results of an EDMA comparison of the Ts65Dn aneuploid mandibles with their normal littermates shown on a scatter plot. The Y-axis represents the value of the elements of the form difference matrix while landmark numbers are shown along the X-axis. The ratio value for the comparison of each linear distance is shown twice on the graph, once for each landmark that is an endpoint for the linear distance. Figure 4.11a shows that the largest range of ratio values are associated with linear distances that have landmark 1 as an endpoint. In Figure 4.11b all ratios associated with linear distances that have landmark 1 as an endpoint are graphed as closed diamonds showing that the outlying values (in either the direction greater than or less than 1) for each landmark map to the distance between that landmark and landmark 1.

4.13 Summary

This chapter has provided the basic tools for comparing forms using EDMA. We have provided explanations for the inadequacy of superimposition and deformation approaches to the comparison of forms. We have provided an invariant approach to the statistical analysis of biological forms, along with the analysis of example data sets to illustrate the nuances of our approach. We have underscored the importance of careful application of our methods and thoughtful interpretation of the results. Part 2 of this chapter presents the statistical theory and the computational algorithms for the comparison of forms. Chapter 5 presents the use of these methods in the study of growth.

Statistical Theory for the Comparison of Two Forms

In Chapter 3, we studied the effects of the nuisance parameters of translation, rotation, and reflection on the identifiability and estimation of the mean form and the covariance parameters. In particular, it was shown that mean forms are identifiable only up to the orbit to which they belong and not their precise location on the orbit. In this chapter, we study the effect of nuisance parameters on the identifiability and estimation of the differences in the mean forms of two populations.

The statistical setting is as follows. We have two populations under study. Following the convention of the last chapter we assume that they follow matrix valued Gaussian distributions. Let $X_1, X_2, ..., X_n$ be the landmark coordinate matrices for the sample of n individuals from the first population. Let us assume that $X_i \sim N(M_1 R_i + 1 t_i, \Sigma_{K,1}, R_i^T \Sigma_{D,1} R_i)$ for $i = 1, 2, ..., n$. Similarly, let $Y_1, Y_2, ..., Y_m$ be the landmark coordinate matrices for the sample of m individuals from the second population. Let us assume that $Y_j \sim N(M_2 R_j + 1 t_j, \Sigma_{K,2}, R_j^T \Sigma_{D,2} R_j)$ for $j = 1, 2, ..., m$.

When comparing two forms, the interest usually lies in studying the relationship between the two mean forms M_1 and M_2. Recall that mean forms are identifiable only up to the orbit to which they belong and the precise location on this orbit cannot be specified. An implication of this result is that we can estimate those differences between the mean forms that depend only on the specification of the orbits and that we cannot estimate differences that require information as to a specific location on the orbit. Hence, we make the following definition.

Definition 1: Let *Diff* (M_1, M_2) be a measure of the difference between two mean forms M_1 and M_2. Then *Diff* (M_1, M_2) is an invariant measure of form difference if and only if *Diff* $(M_1, M_2) = $ *Diff* $(M_1 R_1 + 1 t_1^T, M_2 R_2 + 1 t_2^T)$ for all orthogonal matrices R_i and translation vectors t_i.

In other words, the notion of form difference should not depend on which members of the two orbits, corresponding to M_1 and M_2, are chosen to calculate the form difference. One should be able to take any member of the first orbit, any member of the second orbit, and calculate a form difference that is invariant to the choice of the particular members.

There are several different measures of form difference that are used in practice. We examine some of the prominent ones in light of the above definition.

4.14 Deformation approach to form difference

Let M_1 and M_2 be two mean forms under consideration. A typical way to represent the relationship between M_1 and M_2 is to consider M_1 being deformed into M_2 by a function h. Typically a parametric family of functions represents this deformation. Thus we have $M_2 = h(M_1, \phi)$ and the value of the parameter ϕ helps us understand the underlying biological mechanisms that are associated with this deformation.

To make the basic ideas understandable, let us consider the simplest case of two-dimensional objects with three landmarks each. We assume that the only possible deformation is of the simplest type — an affine deformation. Under this model, $M_2 = M_1 A + \underline{1}t^T$ where A is any 2 x 2 matrix and t is a translation vector. Thus the parameter of interest is A and t is the nuisance parameter. It is easy to see that $A = (LM_1)^{-1}(LM_2)$ where L is the translation matrix defined in Chapter 3. A physical interpretation can be given to this matrix A by writing its singular value decomposition in the following manner (Goodall and Green, 1986):

$$A = \begin{pmatrix} \cos\theta & -\sin\theta \\ \sin\theta & \cos\theta \end{pmatrix} \begin{pmatrix} p & 0 \\ 0 & q \end{pmatrix} \begin{pmatrix} \cos\varphi & -\sin\varphi \\ \sin\varphi & \cos\varphi \end{pmatrix}$$

The angle θ corresponds to the initial rotation to the principal axes of M_1, the singular values p and q correspond to the stretching along the principal axes and the angle φ corresponds to the rotation of the deformed M_1 to match exactly with M_2. From the scientific point of view, the parameters of interest are (θ, p, q) because they contribute to the deformation, whereas the parameter φ does not. Let us examine whether A is invariant as a measure of form difference. In the notation used in definition 1, $Diff(M_1, M_2) = A = (LM_1)^{-1}(LM_2)$, whereas, $Diff(M_1 R_1 + \underline{1}t_1^T, M_2, R_2 + \underline{1}t_2^T) = R_1^{-1}AR_2 = R_1^{-1}Diff(M_1, M_2)R_2$. Thus, this measure of form difference is not invariant. It depends on which members of the orbits are chosen to evaluate the affine deformation. However, if one is interested in the singular values of the affine deformation matrix A, then it is clear that singular values are invariant. Thus, the measure of anisotropy, defined as the logarithm of the ratio of the singular values (Small, 1996, page 95), is invariant.

The affine deformation is the simplest type of deformation function. In practice, more complex types of deformation functions are required. For example, Bookstein (1989, 1991, and 1996) suggests the use of thin plate splines as a deformation function. An advantage of thin plate splines is that it depicts the deformation geometrically and that the total deformation can be decomposed into several orthogonal components which, in turn, can be studied in order to localize the form differences. Such localization, presumably, helps understand the biological processes that might be responsible for the form change. We refer the reader to Bookstein's work for more details and applications. We would also like to point out that deformation functions other than the thin-plate splines may also be used.

We now consider the non-invariance of the thin plate splines in more detail. Assume that the objects under consideration are two-dimensional. Let d_{ij} denote the Euclidean distance between landmarks i and j in M_1, and let $U(d) = d^2 \log d^2$. Let U be a $K \times K$ matrix given by $U = [U(d_{ij})]_{i,j=1,2,\ldots,K}$. According to the thin plate splines model, the relationship between the two forms is given by: $M_2 = 1t^T + M_1 A + UN$ where t is the translation parameter, A is a 2 x 2 matrix corresponding to the affine deformation, and N is a $K \times 2$ matrix corresponding to the non-affine deformation parameters. Notice that there are $(2K + 4 + 2)$ unknown parameters whereas the system of equations given by $M_2 = 1t + M_1 A + UN$ has only $2K$ equations. In order to make the parameters identifiable, one needs additional constraints. Traditionally, in the thin-plate splines literature (e.g.,Cressie, 1991, page 181), additional constraints are imposed by presuming that the parameter N is such

that $\begin{bmatrix} 1^T \\ M_1^T \end{bmatrix} N = 0$. The resultant thin-spline function is termed as the

'minimum bending energy' spline.

There are, of course, any number of different external constraints that may be considered to make the parameters identifiable. A class of such constraints that is considered in practice (Rohlf, 1996) can be

described as follows. Let $V = \begin{bmatrix} 1^T \\ M_1^T \end{bmatrix}$ and W be any $K \times K$ symmetric matrix. The constraint specifies that the parameters corresponding to translation and affine deformation are such that

$$(VWV^T)^{-1}VWM_2^T = \begin{bmatrix} t^T \\ A \end{bmatrix} \quad \text{and then} \quad U^{-1}(M_2 - 1t^T - M_1A) = N.$$

Depending on the choice of the weight matrix, one obtains different solutions.

An important point to make is that this choice is external to the group structure. Fixing this constraint is analogous to fixing a linear constraint among the α_i in the model for a one-way ANOVA: $E[y_{ij}] = \mu + \alpha_i$. For example, Arnold (1981) considers the constraint $\sum_{i=1}^{I} \alpha_i = 0$. One can also use the constraint $\alpha_1 = 0$ or any number of other constraints. For the sake of estimation, such an arbitrary constraint may be necessary, however when it comes to scientific and statistical inference, such a constraint should have no effect. For example, in the analysis of variance, inferences are based only on estimable functions of the α_i's, which are invariant to the choice of the constraint.

Analogous to the linear models situation, the principle of invariance suggests that such external constraints should have no effect on the statistical inference when the deformation approach is used to define form difference. Unfortunately the thin-plate methodology does not satisfy the principle of invariance. Such lack of invariance is also unappealing from a scientific point of view. Those inferences that vary depending on the choice of the external constraints are, generally, scientifically unacceptable. In Part 1, Section 5 of this chapter, we have illustrated the scientific consequences of this lack of invariance of the thin-plate splines methodology.

4.15 Superimposition methods for comparison of forms

We now examine the superimposition method for comparing two forms. For details on this method, we refer the reader to Siegel and Benson (1982), Rohlf and Slice (1990) and Lele (1991). This method of form difference analysis has been in use in morphometrics since the beginning of this century (Cole, 1996) and has been recently proposed by Goodall (1991) and Rohlf and Slice (1990). Dryden and Mardia (1998) provide a book length treatment of this approach. Small (1996, page 35) gives a succinct description of this approach as "an attempt to discover the shape differences between sets will typically involve a matching of the sets to determine how differences in the coordinates of corresponding points can be explained through similarity transformations. Any residual differences that cannot be explained through similarity transformations can be understood to be due to differences in shape."

The model used in the superimposition approach may be described in mathematical terms as follows. Suppose we are interested in describing the difference in two forms M_1 and M_2. The superimposition approach postulates that the two forms are related to each other by the model: $M_2 = 1 t^T + M_1 R + D_F$ where t is a translation parameter, R is an orthogonal matrix and the matrix D_F describes the form difference. If one is interested in shape differences, the model is given by $M_2 = 1 t^T + \alpha M_1 R + D_S$ where $\alpha > 0$ is a real number, taking into consideration the scaling of an object and the residual matrix D_S describes the shape difference. The model component $1 t^T + \alpha M_1 R$ is known as the similarity transformation because the resultant objects after this transformation are geometrically similar to the original object M_1. According to the superimposition approach, what remains after the similarity transformation, namely D_S, describes the shape difference.

Consider the model describing the form difference: $M_2 = 1 t^T + M_1 R + D_F$. Given M_1 and M_2, we want to determine t, R and D_F such that this equation is satisfied. Suppose the objects under consideration are two-dimensional. Then, there are 2 parameters related to translation — one parameter related to the angle of rotation and $2K$ parameters related to the form difference. It can be noted immediately that there are $2K + 3$ unknowns and $2K$ equations, hence, there is no unique solution. To make the form difference identifiable, superimposition methods put additional constraints on these parameters. For example, one Procrustes superimposition approach (Siegel and Benson, 1982), also known as the least squares fitting criterion, applies the constraint that the rotation and translation parameters are such that $tr\{(M_2 - M_1 R - 1 t)(M_2 - M_1 R - 1 t)^T\}$ is minimum. Once the translation and rotation parameters are fixed in this fashion, the form difference is well defined. There are other sets of constraints that are also used in practice. For example, consider the edge superimposition scheme commonly used in roentgenographic cephalometry. This particular method of studying form difference translates and rotates M_1 in such a manner that an edge joining a chosen pair of landmarks matches with the corresponding edge in M_2. This, in effect, imposes the constraint that one particular row and one other element of D_F is zero.

It is easy to show that the form difference matrix D_F obtained by the least squares constraint will be different than the one obtained by the edge superimposition constraint.

Similar to the discussion of the deformation approach, the principle of invariance suggests that we should restrict ourselves to inferences that are invariant to the choice of the external constraint. In Part 1, Section 5 of this chapter, we have illustrated the practical consequences of the lack of invariance of the superimposition approach. In the following, we discuss the relationship between invariance and identifiability in a mathematically rigorous fashion and show that the transformations considered by the deformation and superimposition approaches are non-identifiable.

4.16 Matrix transformations, invariance, and identifiability issues

The problem of studying form differences can be put in terms of studying transformation of one $K \times D$ matrix into another $K \times D$ matrix. The invariance ideas described in Chapter 3 can then be easily extended to the two-sample problem. We then discuss the non-identifiability of the transformations considered by the superimposition and deformation approaches. Our argument is that non-identifiable transformations should not be used for statistical or scientific inferences. We then show that identifiability is equivalent to invariance. In the next section we discuss invariant procedures to study form difference.

Consider the class of all transformations that map the space of $K \times D$ matrices to the space of $K \times D$ matrices. Let Ω denote the space of all $K \times D$ matrices. Let $H : \Omega \to \Omega$ be the collection of all transformations that map any given $K \times D$ matrix to another $K \times D$ matrix. For example, the affine transformation discussed earlier belongs to this class, so do the thin-plate splines and superimposition transformations belong to this class.

Recall that two $K \times D$ matrices are equivalent to each other if they are rotations, reflections, and/or translations of each other. We can partition the space Ω into equivalency classes, each class consisting of equivalent matrices. Let $O(M)$ denote the class of matrices equivalent to a given matrix M.

Let h_1 and h_2 be two members of H such that $h_1(MR_1 + \underline{1}t_1) = h_2 (MR_2 + \underline{1}t_2)R + \underline{1}t$. That is, the resultant matrices from both transformations acting on equivalent matrices are also equivalent to each other.

We say that these two transformations are equivalent to each other. Let us denote by H_{M_1,M_2}, the set of transformations that map an element of the orbit of M_1 to an element of the orbit of M_2.

The key result that follows from the maximal invariance machinery developed in Chapter 3, Part 2, is that:

For $h_1, h_2 \in H_{M_1,M_2}$, $likelihood(h_1|Data) = likelihood(h_2|Data)$.

That is, data cannot help us distinguish between the transformations that map an element of the orbit of M_1 to an element of the orbit of M_2. It is thus imperative that our inferences be invariant to the choice of a particular transformation within H_{M_1,M_2}. This is precisely the requirement specified in definition 1.

It seems, however, that the superimposition and deformation methods are able to specify a particular transformation from this class. How is that possible? The superimposition methods choose one of the transformations H_{M_1,M_2} using an external criterion such as the least squares. Similarly the thin-plate splines methods choose a particular transformation that minimizes the bending energy or some such external criterion. *No amount of data can instruct us which of these external criterion is valid in nature.* We have shown in Part 1 of this chapter that the inferences based on these different criteria can be quite distinct from each other. Our argument is that since data cannot distinguish between these different criteria and hence these different scientific inferences, the proper thing to do is to base statistical and scientific inferences *only* on those quantities that remain invariant to the choice of a particular transformation in H_{M_1,M_2}.

To argue that this may be problematic, we consider a somewhat simpler situation where the issues are transparent. Suppose we are studying two treatments for reducing blood pressure. Suppose we have several physicians in the study. Each physician selects one of his patients and administers one of the treatments. After a month, reduction in the blood pressure is reported as reduction in diastolic and systolic blood pressure.

Let $X_1, X_2, \ldots, X_n$ denote the observations under the first treatment and $Y_1, Y_2, \ldots, Y_m$ denote the observations under the second treatment. Suppose we model these data as: $X_i \sim N(\mu + \underline{1}t_i, I)$ and $Y_j \sim N(\mu + \Delta + \underline{1}t_j, I)$ where $\mu = \begin{pmatrix} \mu_1 \\ \mu_2 \end{pmatrix}$ is the mean reduction in the diastolic and systolic blood pressure due to treatment 1, $\Delta = \begin{pmatrix} \Delta_1 \\ \Delta_2 \end{pmatrix}$ is the effect of treatment 2 over and above treatment 1 and t_i's correspond to the individual physician effects.

This problem belongs to the class of problems considered by Neyman and Scott (1948). Suppose we use invariance to eliminate the nuisance parameters t_i's corresponding to the physician effect. Notice that $X_{i,2}-X_{i,1}$ is a maximal invariant under the group of translations and $X_{i,2}-X_{i,1} \sim N((\mu_2-\mu_1),2)$. Similarly $Y_{j,2}-Y_{j,1} \sim N((\mu_2-\mu_1)+(\Delta_2-\Delta_1),2)$.

Let us denote $\mu_2-\mu_1=d_\mu, (\Delta_2-\Delta_1)=d_\Delta$. Now consider the partition of the original parameter space $\Theta=\{(\mu,\Delta,\Sigma):\mu\in R^2, \Delta\in R^2\}$ induced by this maximal invariant: $Orbit(d_\mu,d_\Delta)=\{(\mu,\Delta)\in\Theta$ such that $\mu_2-\mu_1=d_\mu,(\Delta_2-\Delta_1)=d_\Delta\}$.

Now consider the likelihood for two sets of parameters, (μ^*,Δ^*) and $(\tilde{\mu},\tilde{\Delta})$ where $\mu_2^*-\mu_1^*=\tilde{\mu}_2-\tilde{\mu}_1$ and $\Delta_2^*-\Delta_1^*=\tilde{\Delta}_2-\tilde{\Delta}_1$, that is, they belong to the same orbit induced by the maximal invariant. Since the distribution of the maximal invariant is a function of (d_μ,d_Δ), it is clear that the two combinations of the parameters have identical likelihood. An immediate consequence of this is that Δ_1 and Δ_2, the change in the diastolic blood pressure and the change in the systolic blood pressure, are not identifiable. We can only find out if the change in the diastolic blood pressure is larger or smaller than the change in the systolic blood pressure and the magnitude of that difference. But the question: How much does the treatment 2 reduce systolic blood pressure or diastolic blood pressure, is unanswerable in this situation.

Instead of accepting this limitation of the data, we decide to follow the superimposition logic and conduct the following analysis.

1. **Least squares approach:** Since there are nuisance parameters related to the translation group, we first superimpose, that is translate, the observations in Sample 1 in such a manner that $\sum_{i=1}^{n}\sum_{j>i} tr\{(X_i-X_j)(X_i-X_j)^T\}$ is minimized. This corresponds to transforming the observations $X_i = \begin{pmatrix} X_{i1} \\ X_{i2} \end{pmatrix}$

to $\tilde{X}_i = \begin{pmatrix} \dfrac{X_{i1}-X_{i2}}{2} \\ \dfrac{X_{i2}-X_{i1}}{2} \end{pmatrix}$. Similarly, we translate the observations

in Sample 2 to $\tilde{Y}_i = \begin{pmatrix} \dfrac{Y_{i1}-Y_{i2}}{2} \\ \dfrac{Y_{i2}-Y_{i1}}{2} \end{pmatrix}$. Now, we average these transformed observations to obtain $\overline{\overline{X}}$ and $\overline{\overline{Y}}$. It is easy to see that

$$\overline{\overline{X}} \sim N\left(\begin{pmatrix} (\mu_1 - \mu_2)/2 \\ (\mu_2 - \mu_1)/2 \end{pmatrix}, \frac{1}{n}\begin{pmatrix} 0.5 & -0.5 \\ -0.5 & 0.5 \end{pmatrix}\right) \text{ and}$$

$$\overline{\overline{Y}} \sim N\left(\begin{pmatrix} (\mu_1 - \mu_2)/2 \\ (\mu_2 - \mu_1)/2 \end{pmatrix} + \begin{pmatrix} (\Delta_1 - \Delta_2)/2 \\ (\Delta_2 - \Delta_1)/2 \end{pmatrix}, \frac{1}{m}\begin{pmatrix} 0.5 & -0.5 \\ -0.5 & 0.5 \end{pmatrix}\right).$$

To study the difference between the two populations, now we superimpose, that is translate, $\overline{\overline{X}}$ so that it matches $\overline{\overline{Y}}$ in the sense that $tr\{(\overline{\overline{X}} - \overline{\overline{Y}})(\overline{\overline{X}} - \overline{\overline{Y}})^T\}$ is minimized. It can be seen that the best translation in this case is no translation at all, and that the difference between two populations is given by: $\overline{\overline{Y}} - \overline{\overline{X}}$. As the sample sizes converge to infinity, this quantity converges to $\begin{pmatrix} (\Delta_1 - \Delta_2)/2 \\ (\Delta_2 - \Delta_1)/2 \end{pmatrix} = \begin{pmatrix} -\delta_\Delta/2 \\ \delta_\Delta/2 \end{pmatrix}$.

2. **Edge superimposition analysis:** We can conduct similar calculations where instead of translating the observations so that least squares criterion is satisfied, we translate them so that the first component (landmark 1) is matched. Such calculations lead to the difference in two populations which converges to $\begin{pmatrix} 0 \\ (\Delta_2 - \Delta_1) \end{pmatrix} = \begin{pmatrix} 0 \\ \delta_\Delta \end{pmatrix}$. It leads us to the conclusion that treatment 2 is effective only in reducing the systolic blood pressure and has no effect on the diastolic blood pressure. Had we decided to match the second component (landmark 2), we would have come to the conclusion that treatment 2 only affects the systolic blood pressure and not the diastolic blood pressure.

It should be obvious from the above example, that the choice of the side conditions has a significant impact on the scientific conclusions and that even an infinite amount of data cannot tell us which side condition is "correct." One may argue that the choice of the side condition is similar to the choice of the "Normal" distribution model. The difference between choosing a particular model such as the Normal distribution in the above example and choosing a side condition is that data potentially can refute or support the choice of the Normal model. However, there is no potential for refuting or supporting a particular choice of the side condition.

The situation in morphometrics is the same as above, except that the group structure is more complicated, involving rotations and reflections. Since the choice of the side condition is not even potentially refutable, we suggest that only those conclusions that remain invariant to the choice of these side conditions are valid conclusions. Immediate questions that come to mind are: what quantities do remain invariant to the choice of the side conditions, and are they of any use in the scientific inquiries? We address the first question here. The second issue of scientific relevance is addressed in Part 1 of this chapter.

4.17 Form comparisons based on distances

In the first part of this chapter, we have described various notions of the form difference and shape difference matrix which consist of the ratios of the corresponding distances in the two mean forms. Let us look at one particular definition of form difference, the form difference matrix, from the invariance perspective. Other notions, such as the shape difference matrix, can be studied in a similar fashion.

Recall that the form difference matrix consists of the ratios of like linear distances in the two forms, namely,

$$FDM(M_1, M_2) = \left[\frac{FM_{ij}(M_1)}{FM_{ij}(M_2)} \right].$$ Now suppose we rotate, reflect and

translate the two forms, each one differently, and then calculate the form difference matrix. We know that rotation, translation and reflection of an object does not affect the form matrix, that is, $FM(M) = FM(MR + 1t)$.

$$FDM(M_1 R_1 + 1t_1, M_2 R_2 + 1t_2) = \left[\frac{FM_{ij}(M_1 R_1 + 1t_1)}{FM_{ij}(M_2 R_2 + 1t_2)} \right] = \left[\frac{FM_{ij}(M_1)}{FM_{ij}(M_2)} \right].$$

It is clear that $FDM(M_1, M_2) = FDM(M_1 R_1 + 1t_1{}^T, M_2 R_2 + 1t_2{}^T)$. This measure of form difference, thus, satisfies definition 1.

Let us now study some important mathematical properties of the form matrix and related quantities.

4.18 Form space based on Euclidean Distance Matrix representation

We noted in Chapter 3 that every object in D-dimensional space represented by K landmarks corresponds to some $K \times D$ matrix, and, conversely, it is also true that corresponding to every $K \times D$ matrix, there exists an object in D-dimensional space represented by K landmarks.

Let X be any $K \times D$ matrix. Let $FM(X)$ be the form matrix corresponding to X. Recall that the form matrix is a square, symmetric matrix with zeros along the diagonals and consists of all pairwise inter-landmark distances in the object X. Let us collect all the upper diagonal entries of the form matrix in a vector of length $K(K-1)/2$. It is easy to see that this vector can be represented as a point in the Euclidean space of dimension $K(K-1)/2$. Thus, every object in D-dimensional space represented by K landmarks can also be represented by a single point in a Euclidean space of dimension $K(K-1)/2$.

Now, conversely, suppose that we are given an arbitrary point in the Euclidean space of dimension $K(K-1)/2$. When does such a point correspond to some D dimensional object with K landmarks?

One condition is obvious because the distances are always positive, such a point should belong to the positive quadrant of the Euclidean space. However, this is not a sufficient condition. In Part 1 of this chapter, we noticed that corresponding to a collection of three positive numbers or equivalently, a point in the positive quadrant of a three dimensional Euclidean space, there corresponds a triangle or a three landmark object in two dimensions, if and only if the sum of any two numbers exceeds the third number. The following discussion generalizes this result to the case when we have K landmarks and D dimensional objects. We first introduce some additional notation and then state the necessary and sufficient conditions under which a point in the positive quadrant of a Euclidean space of dimension $K(K-1)/2$ corresponds to a D-dimensional object represented by K landmarks.

Let $\underline{d}$ be a point in the positive quadrant of the Euclidean space of dimension $K(K-1)/2$. Let $\underline{d}^T = (d_{12}, d_{13}, \ldots, d_{1K}, d_{23}, d_{24}, \ldots, d_{2K}, \ldots d_{(K-1)K})$ be the elements of this vector. Construct two new matrices using the individual elements of this vector:

$$D_d = \begin{bmatrix} 0 & d_{12} & d_{13} & \cdots & d_{1K} \\ d_{12} & 0 & d_{23} & \cdots & d_{2K} \\ d_{13} & d_{23} & 0 & \cdots & \vdots \\ \vdots & \vdots & \vdots & 0 & d_{(K-1)K} \\ d_{1K} & \cdots & \cdots & d_{(K-1)K} & 0 \end{bmatrix}, \text{ and,}$$

$$E_d = \begin{bmatrix} 0 & d_{12}^{\,2} & d_{13}^{\,2} & \cdots & d_{1K}^{\,2} \\ d_{12}^{\,2} & 0 & d_{23}^{\,2} & \cdots & d_{2K}^{\,2} \\ d_{13}^{\,2} & d_{23}^{\,2} & 0 & \cdots & \vdots \\ \vdots & \vdots & \vdots & 0 & d_{(K-1)K}^{\,2} \\ d_{1K}^{\,2} & \cdots & \cdots & d_{(K-1)K}^{\,2} & 0 \end{bmatrix}.$$

$$\text{Let} \quad H = \begin{bmatrix} 1-\dfrac{1}{K} & -\dfrac{1}{K} & \cdots & -\dfrac{1}{K} \\ -\dfrac{1}{K} & 1-\dfrac{1}{K} & \cdots & \vdots \\ \vdots & \vdots & \vdots & \vdots \\ -\dfrac{1}{K} & -\dfrac{1}{K} & \cdots & 1-\dfrac{1}{K} \end{bmatrix}$$

and define a new matrix, $B_d = \dfrac{-1}{2} H E_d H^T$.

Theorem 1: Corresponding to a point $\underline{d} \in E^{K(K-1)/2}$ there exists a D-dimensional object with K landmarks, if and only if:

i) $\underline{d}$ belongs to the positive quadrant, that is, all the elements of this vector are non-negative, and,

ii) The matrix B_d is a positive semi-definite matrix with $rank(B_d)=D$.

Proof: This is a standard result in multidimensional scaling literature. See Mardia et al. (1979, Chapter 14) for details.

Similar to the discussion about the form space corresponding to three-landmark, two-dimensional objects in Part 1, it can be noticed that the set of points that satisfies the above condition constitute a subset of the positive quadrant of the Euclidean space of dimension

$K(K-1)/2$. The above theorem provides a characterization of this subset of the $K(K-1)/2$ dimensional Euclidean space that corresponds to forms of K-landmark, D-dimensional objects. We call this subset "the form space of K-landmarks, D-dimensional objects" or "the form space," for short.

Suppose now that we are given a point $\underline{d}$ from the form space. How do we obtain the landmark coordinates of the corresponding object? It is clear that the mapping from the form space to the space of landmark coordinate matrices is not one-to-one. A single point in the form space maps onto a single orbit, which is, in fact, a collection of several matrices, in the space of landmark coordinate matrices. Thus, corresponding to a single point in the form space, there is a unique orbit in the landmark coordinate matrix space and conversely, corresponding to a given orbit. there is a unique point in the form space. For a discussion of orbits or equivalence classes, see Chapter 3 or the first part of this chapter. In the following, we show how to map a given point in the form space to a particular member of the corresponding orbit. Once such a member is obtained, one can rotate, reflect, or translate it to obtain any other member in the orbit.

Let $\underline{d}$ be a point in the form space. Calculate B_d. Consider the spectral decomposition of the matrix B_d, namely, $B_d = P\Lambda P^T$. Because the rank of matrix B_d is D and is positive semi-definite, there are D nonzero eigenvalues. Let $\lambda_1, \lambda_2, \ldots, \lambda_D$ be the eigenvalues and let $P_{(1)}, P_{(2)}, \ldots, P_{(D)}$ be the corresponding eigenvectors. Construct a $K \times D$ matrix as follows: $A_d = \left[\sqrt{\lambda_1} P_{(1)} \quad \sqrt{\lambda_2} P_{(2)} \quad \cdots \quad \sqrt{\lambda_D} P_{(D)} \right]$. This matrix is a landmark coordinate representation of the object corresponding to $\underline{d}$. If we calculate the form matrix corresponding to A_d, namely the matrix of all possible pairwise distances between these landmark coordinates, it exactly equals $\underline{d}$.

The results above provide a way to transform a landmark coordinate representation to the form matrix representation and vice versa.

4.19 Statistical properties of the estimators of mean form, mean form difference, and mean shape difference matrices

Under the Gaussian perturbation model described in Chapter 3, we show that the mean form difference and mean shape difference matrices can be estimated consistently.

Theorem 2: Let $X_1, X_2, \ldots, X_n$ be independent, identically distributed, matrix random variables following the distribution $N(M, \Sigma_K, I)$. As the sample size n increases, $FM(\hat{M}) \to FM(M)$. That is, the mean form matrix estimator described in Chapter 3 is consistent. Similarly $\hat{\Sigma}_K^{\bullet} \to L\Sigma_K L^T$.

Proof: The results follow from the fact that the sample moments are consistent estimators of the population moments and that $FM(M)$ and $L\Sigma_K L^T$ are continuous functions of sample moments and hence are also consistent.

Theorem 3: Let $X_1, X_2, \ldots, X_n$ be independent, identically distributed, matrix random variables following the distribution $N(M_1, \Sigma_{K,1}, I)$. Similarly let $Y_1, Y_2, \ldots, Y_m$ be independent, identically distributed, matrix random variables following the distribution $N(M_2, \Sigma_{K,2}, I)$. Let samples X and Y be independent of each other. As the sample sizes n and m increase,

$$FM(\hat{M}_2, \hat{M}_1) \to FM(M_2, M_1).$$

Proof: This follows from Theorem 2 and the fact that $FDM(M_2, M_1)$ is a continuous function of $FM(M_1)$ and $FM(M_2)$.

Similar results can be proved in the case where the data arise from the distribution $N(M, \Sigma_K, \Sigma_D)$. The details are similar and not presented here.

4.20 Computational algorithms

In this part, we describe the computer algorithms that are used for the bootstrap and Monte Carlo testing procedures and confidence intervals for mean form difference matrix.

Bootstrap-based procedures

Bootstrap procedures may be used provided the sample size is larger than 15 or 20. The main advantage of these procedures is that they do not assume that the perturbations are Gaussian random variables.

EDMA-I testing procedure for testing equality of shapes

As explained in Part 1 of this chapter, the EDMA-I testing procedure requires a choice of one of the populations as the reference or the base population. The null hypothesis to be tested is that the mean of the second population is a scaled version of the mean form of the base population. More precisely, the question being answered is similar to the hypothesis in a classification problem: Could the sample from the second population have arisen from the first population? In order to apply this testing procedure we do not need to assume that the perturbations are Gaussian. However, we do need to assume that the covariance matrices for the two populations are proportional to each other.

STEP 1: Calculate the form difference matrix using the estimated mean forms for the two populations and calculate

$$T_{obs} = \frac{\max_{ij} FDM_{ij}(\hat{M}_1, \hat{M}_2)}{\min_{ij} FDM_{ij}(\hat{M}_1, \hat{M}_2)}.$$

STEP 2: Generate new set of samples $X_1^*, X_2^*, \ldots, X_n^*$ and $Y_1^*, Y_2^*, \ldots, Y_m^*$ with replacement from the base or the reference sample.

STEP 3: Calculate $\quad T^* = \dfrac{\max_{ij} FDM_{ij}(\hat{M}_1^*, \hat{M}_2^*)}{\min_{ij} FDM_{ij}(\hat{M}_1^*, \hat{M}_2^*)}.$

STEP 4: Repeat Steps 2 and 3 B times to obtain $T_1^*, T_2^*, \ldots, T_B^*$.
STEP 5: Arrange $T_1^*, T_2^*, \ldots, T_B^*$ in an increasing order. Reject the null hypothesis of equality of shapes if T_{obs} is in the top α-th percentile where α is the size of the test.

Confidence intervals for the form difference matrix

The Bootstrap procedure to obtain confidence intervals for the form difference matrix does not need the choice of a reference or base population, nor does it require the assumption of proportional covariance matrices or a Gaussian perturbation model. The algorithm for computing bootstrap confidence intervals is detailed below.

STEP 1: Calculate the form difference matrix using the estimated mean forms for the two populations.

STEP 2: Generate a new set of samples $X_1^*, X_2^*, \ldots, X_n^*$ with replacement from the first sample $X_1, X_2, \ldots, X_n$ and $Y_1^*, Y_2^*, \ldots, Y_m^*$ with replacement from the second sample $Y_1, Y_2, \ldots, Y_m$.

STEP 3: Calculate the form difference matrix based on these bootstrap samples.

STEP 4: Repeat Steps 2 and 3 B number of times.

The bootstrap sample based form difference matrices obtained in this fashion can be collected in a matrix with $K(K-1)/2$ rows and B columns. In this matrix, each column is a form difference matrix obtained at the end of Step 3 and each row represents B form difference ratios for a linear distance between a specified pair of landmarks.

To obtain a confidence interval for each linear distance, the ratios in each row are sorted in an increasing order. If a 90% confidence interval is sought, then the first 5% and the last 5% of the total entries in this sorted row are deleted. The minimum and the maximum entries remaining in that row constitute the lower and upper confidence limits for that particular linear distance. This is done separately for each row to obtain a confidence interval for each linear distance ratio.

Monte Carlo-based procedures

The difference between the Bootstrap procedures described above and the Monte Carlo procedures described below is that in the Monte Carlo procedure, instead of obtaining Bootstrap samples using simple random sampling with replacement from the original sample, we generate Bootstrap samples using the Gaussian model and the estimated mean form and the variance-covariance matrix. This procedure is valid provided the Gaussian perturbation model is a reasonable model. This procedure may be used when the sample sizes are small, say 10-15 individuals in each group.

Monte Carlo confidence intervals for the form difference matrix

STEP 1: Calculate the form difference matrix using the estimated mean forms for the two populations.

STEP 2: Generate a new set of samples $X_1^*, X_2^*, \ldots, X_n^*$ from $N(\hat{M}_1, \hat{\Sigma}_{K,1}, I)$ and $Y_1^*, Y_2^*, \ldots, Y_m^*$ from $N(\hat{M}_2, \hat{\Sigma}_{K,2}, I)$.

STEP 3: Calculate the form difference matrix based on these Monte Carlo samples.
STEP 4: Repeat Steps 2 and 3 B number of times.

The Bootstrap sample based form difference matrices obtained in this fashion can be collected in a matrix with $K(K-1)/2$ rows and B columns. In this matrix each column is a form difference matrix obtained at the end of Step 3 and each row represents B form difference ratios for a linear distance between a specified pair of landmarks.

To obtain a confidence interval for each linear distance, the ratios in each row are sorted in an increasing order. If a 90% confidence interval is sought, then the first 5% and the last 5% of the total entries in this sorted row are deleted. The minimum and the maximum entries remaining in that row constitute the lower and upper confidence limits for that particular linear distance. This is done separately for each row to obtain a confidence interval for each linear distance ratio.

Monte Carlo-based EDMA-I testing procedure

STEP 1: Calculate the form difference matrix using the estimated mean forms for the two populations. Calculate

$$T_{obs} = \frac{\max_{ij} FDM_{ij}(\hat{M}_1, \hat{M}_2)}{\min_{ij} FDM_{ij}(\hat{M}_1, \hat{M}_2)}.$$

STEP 2: Assuming that the first population is the base sample, generate a new set of samples $X_1^*, X_2^*, ..., X_n^*$ from $N(\hat{M}_1, \hat{\Sigma}_{K,1}, I)$ and $Y_1^*, Y_2^*, ..., Y_m^*$ from $N(\hat{M}_1, \hat{\Sigma}_{K,1}, I)$.

STEP 3: Calculate $\quad T^* = \dfrac{\max_{ij} FDM_{ij}(M_1^*, M_2^*)}{\min_{ij} FDM_{ij}(M_1^*, M_2^*)}.$

STEP 4: Repeat Steps 2 and 3 B times to obtain $T_1^*, T_2^*, ..., T_B^*$.
STEP 5: Arrange $T_1^*, T_2^*, ..., T_B^*$ in an increasing order. Reject the null hypothesis of equality of shapes if T_{obs} is in the top α-th percentile where α is the size of the test.

Monte Carlo-based EDMA-II testing procedure

The difference between EDMA-I and EDMA-II is described in detail in
Part 1 of this chapter. We do not repeat the discussion here.

> STEP 1: Calculate the shape difference matrix using the esti-
> mated mean forms for the two populations. Calculate the
> extreme value in the shape difference matrix, either negative
> or positive.
>
> STEP 2: Assuming that the first population is the base sam-
> ple, generate a new set of samples $X_1^*, X_2^*, ..., X_n^*$ from
> $N(\hat{M}_1, \hat{\Sigma}_{K,1}, I)$ and $Y_1^*, Y_2^*, ..., Y_m^*$ from $N(\hat{M}_1, \hat{\Sigma}_{K,1}, I)$.
>
> STEP 3: Calculate the shape difference matrix based on these
> samples and calculate the most extreme value, negative or
> positive and call it T^*.
>
> STEP 4: Repeat Steps 2 and 3 B times to obtain $T_1^*, T_2^*, ..., T_B^*$.
>
> STEP 5: Arrange $T_1^*, T_2^*, ..., T_B^*$ in an increasing order. Assume
> that the size of the test is 0.10, then calculate the 5-th per-
> centile and the 95-th percentile values of these T^*'s. If zero
> lies in between these values, then accept the null hypothesis
> of equality of shapes, otherwise reject it.

We want to reemphasize the fact that if the distribution of the test
statistic is bimodal, one should not conduct the above test. In general,
we prefer the confidence intervals for the shape or form difference
matrices over simplistic null hypothesis testing.

CHAPTER 5

The Study of Growth

"The form of an organism is determined by its rate of growth in various directions"

D'Arcy Thompson (1992)

Up to this point we have limited our discussion to the description and comparison of two forms. Growth analysis stems directly from the methods we have already developed in the last chapter and is simply a special case of form comparison. Instead of comparing the form of one group to the form of another group, we compare the form of one age class to the form of another age class within the same group. Consequently, the tools that we have presented thus far are adequate for studying growth within a single group.

We define *growth pattern* as the composite of geometric changes in biological structure occurring through ontogenetic time. In many practical situations, we are interested in comparing the growth pattern of one group with the growth pattern of another. The issues of comparing growth patterns lead to different challenges and problems necessitating further development of analytical techniques. This chapter discusses the special difficulties faced in comparing growth patterns, and develops relevant methods. This discussion is based in part on ideas presented in Richtsmeier and Lele (1993).

The study of growth provides important information for many fields of inquiry. In the study of dysmorphology or disease, a valid question concerns the original source and reason for continuation or intensification of the dysmorphology. Is the dysmorphology a product of an initial developmental insult that is carried to a new developmental stage by following a normal growth pattern? Or is growth affected such that the dysmorphology is at least in part a product of an aberrant growth pattern? The answer to this question might inform us about disease process and can be used in the planning of surgical or non-surgical interventions in the case of human malformations (Dufresne and Richtsmeier, 1995).

Studies of growth patterns can provide information about the attainment of features in varying species. Increase in snout length in mammalian species might be attained by growth changes local to the maxillae and premaxillae or by changes in the relative positions and sizes of the pterygoid plates, sphenoidal body or the vomer, as these osseous components determine in part how the face is hafted onto the cranial base. Careful growth analyses can provide information about the timing of developmental events, or changes in magnitudes and rates of growth. This can, in turn, be used to determine the ways in which differences in local growth patterns contribute to difference in morphological traits among species (Richtsmeier, Corner, et al., 1993). The challenge in any analysis of growth is to go beyond quantitative description; it is to use the study of growth to explore aspects of the biology of the organism such as the genetic basis of morphogenesis, phylogenetic underpinnings of developmental patterns, or the role of hormones, teratogens, dietary elements, and other environmental variables on the growth process.

In previous chapters, we discussed in detail the problems relating to registration and lack of a common coordinate system. These problems are equally important when the study concerns growth and the comparison of growth patterns. To give an example, the field of Roentgencephalometry (Broadbent, 1975) was developed to enable quantitative comparison of x-rays. The method was developed to study craniofacial form and growth. It became the method of choice for clinical studies in the 1950s and 1960s and is still commonly used today e.g., (Kreiborg and Pruzansky, 1981; Kreiborg, Aduss, et al., 1999). Figure 5.1 contains a growth series of immature human skulls and shows a fairly standard description of cranial growth based on data from lateral x-rays. According to the method of roentgencephalometry, the center of the pituitary fossa (the landmark sella) is the registration point onto which all cases are superimposed. A line that stretches from sella to the landmark nasion (the intersection of the frontal and nasal bones) is used to define the line of orientation of all x-rays being compared. Following this registration method, all cases are first matched at sella and then rotated such that the orientation of the sella-nasion lines match. This system of registration results in the description and quantification of growth shown in Figure 5.1. This system doesn't allow for growth at sella and doesn't allow for non-linear growth

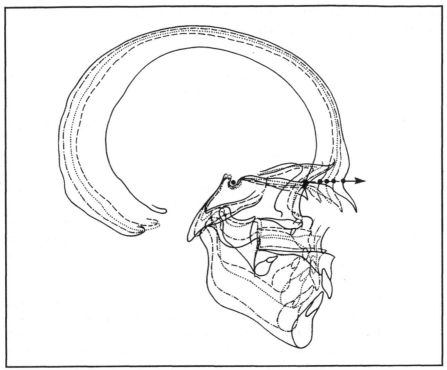

Figure 5.1 Traditional growth series based on longitudinal cephalometric radiographs from the Bolton-Brush growth Series (Broadbent et al., 1975). Tracings of the x-rays are superimposed on the landmark sella located in the middle of the pituitary fossa and oriented on a line that stretches from sella to nasion. The result is an impression of craniofacial growth as an onion skin-like expansion away from the center of the skull marked by sella.

between sella and nasion. Common sense indicates that conditions like these cannot be assumed in the study of growth and research has confirmed this (Moyers and Bookstein, 1979; Richtsmeier and Cheverud, 1986). When superimposition or deformation methods are used for studying growth, the same difficulties that were discussed in the comparison of forms (Chapter 4) are encountered. In this chapter we show how the coordinate system-free method of EDMA can be extended to the study of growth and the comparison of growth patterns.

5.1 Limiting factors in studying growth using morphometric approaches

The methods discussed in this chapter use landmark coordinates to study the way in which component structures of an organism rearrange relative to one another as the organism increases in size. This approach differs from more traditional growth curve analyses (see Zeger and Harlow, 1987) where measures that span gross anatomical regions (e.g., crown-rump length, head circumference) or that represent overall body size (e.g., body weight, stature) are plotted across time as curves that summarize growth patterns. When studying growth, the limitations of landmark data need to be recognized. Beyond the limiting factors presented in Chapter 4, we need to acknowledge that growth is a very complex process. The choice of analyzing landmark data as collected from the morphological consequences of this process places limits on our perceptions. For these reasons, we add two more limiting factors to our list:

4. *Identifiability of the locus of growth.* We cannot infer about the way in which the material between landmarks changes during growth from the information given by the landmark coordinates. If chosen intelligently, the landmark data may capture the evidence of growth processes adequately, but our answers will be limited to changes local to landmarks.
5. *Identifiability of the timing of growth events.* Information obtained on timing and rate of growth is totally dependent upon the time points at which we sample data. We can infer little about the timing of specific growth events if we don't have data that window that event adequately.

5.2 Longitudinal vs. cross-sectional data

There are two major data types that can be used to study growth: longitudinal and cross-sectional. Since growth occurs continuously over time, data points are usually sampled from this continuum for analysis. When working with laboratory animals, the increments can be chosen to fit the research design. In this case, the observer's schedule may be the only complicating factor. If a study of human growth is prospective, as in the case of growth data being collected to establish norms or reference data sets (Garza and DeOnis, 1999), sampling points can be chosen to meet the study criteria. However, most of the time, sampling cannot be done according to a predetermined design, and other factors determine the division of this continuum into the increments that are studied. When growth data are collected from museum specimens or from patient records, the researcher may have little control over the intervals of time included in the growth study.

Longitudinal data refer to data points taken from a single individual at specific intervals over a period of time. Longitudinal data from numerous individuals can be combined to create samples of longitudinal data. Cross-sectional data refer to data taken at specified ages, but

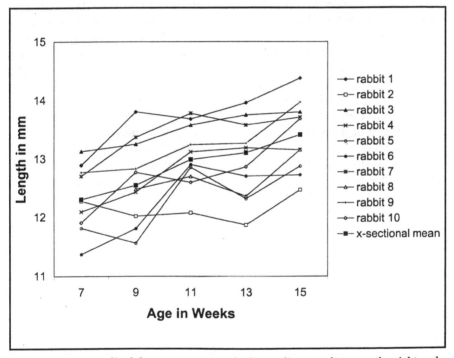

Figure 5.2 Longitudinal data representing the linear distances between the right and left foramen lacerum on the cranial base of the New Zealand white rabbit. Note the variability in the magnitude of growth for each age interval between rabbits. Note also that the length of this distance decreases with age for some rabbits and some age intervals. We have found this to be a normal occurrence during craniofacial growth when studying individual linear distances. An average linear distance was calculated for each age group and plotted as the cross-sectional mean. The cross-sectional mean portrays a steady increase in size over time and does not provide information on the variability in individual growth patterns.

from more than one individual such that a different individual represents each time point. *Cross-sectional data* are often collected from a large population, and each individual is assigned to an age category. In this case, data for a developmentally staged or chronologically aged class consists of data from many individuals, and data from any one individual is only used in a single age class. *Mixed longitudinal data* results when longitudinal data are included within a cross-sectional data set.

Longitudinal data are rare, but extremely valuable when available. Longitudinal data must be analyzed as dependent data since the condition of a data point at time $t+1$ is dependent upon the condition of the data point at time t. Individual modulations in growth patterns can only be identified with closely spaced longitudinal data points (Wilson, 1999). Longitudinal data are of particular use in the study of secular trends, and are a requirement for predictive models of development. On the other hand, closely spaced longitudinal data may obscure more general patterns and reveal seemingly erratic, idiosyncratic patterns of individual growth. To study general population patterns, cross-sectional data may be more useful. The research question determines which types of data are most useful and appropriate.

Growth is a phenomenon of the individual, but samples are needed to speak to the statistical aspects of growth patterns. When longitudinal data for many individuals are plotted over the same age interval, it is clear that growth is highly individualized and quite variable (see Figure 5.2). An average might be offered as representative of the pattern, but these averages can obscure the irregularities observed in individual patterns. A study of averages provides useful information, but because the variability noted in individual growth patterns is responsible in part for inter-individual morphological variability, the variability of individual growth patterns warrants investigation.

5.3 Assigning age and forming age-related groups

Statistical analysis of growth calls for analysis of samples of individuals. To do this, individuals need to be placed into age-graded groups or classes according to some criteria. There are no rigid rules for the formulation of age-specific groups for analysis of growth patterns. If chronological ages are known, it may help to survey the available data and look for natural breaks in the age distribution for forming age groups for analysis. If chronological ages are not known, features of the organisms that have a recognized relationship with chronological age (e.g., tooth eruption patterns, overall body size, annual rings, number of coils, molt stage, epiphyseal fusion, etc.) can be used as age-surrogates and as a means for assigning individuals to developmental age classes.

The researcher needs to be aware of any potential ascertainment biases of the data, and the way in which this bias might affect the age distribution and the results of analysis of these data. When museum samples are used, age distribution may be skewed for reasons that have to do with social behavior of that species. (For example, if a certain subgroup of a wild population protects the group and therefore confronts nonmembers, those individuals will most likely be captured and the resulting collection will be composed primarily of individuals in that age/sex/status group. Methods of capture (e.g., size of nets or traps) may also influence sample composition. If patient data are being accessed, the age at which individuals are diagnosed, treated, or discharged may have implications for the nature of the age groups available. Depending upon the particular research situation, grouping may amplify or reduce within-age group variability.

When growth patterns are being compared between groups, care must be taken to ensure that the criteria used to assign developmental ages are consistent and comparable across groups. This can be extremely difficult, especially when comparisons cross species boundaries. Assigning developmental ages from biological markers can affect the degree of variability within age groups, and this can vary from species to species. For example, let us say that we are interested in comparing growth during infancy between human and non-human primates. The human literature can provide a chronological time frame for the developmental period known as "infancy," and similar definitions for the relationship between chronological age and life history intervals are available for certain non-human primate species raised in laboratory settings. The length of these time frames will be different in each species and this warrants consideration. Extreme care must be taken in estimating ages from proxy data when choosing comparable developmental stages for inter-species comparison. Since each research situation is different, there is little general advice that can be offered beyond caution and alertness to these and other potential hazards.

5.4 EDMA applied to the study of growth

As demonstrated previously for the study of form difference, adoption of an arbitrary coordinate system is undesirable when studying growth, as the results obtained can be misleading. When studying growth, we have representative forms from various age groups, and the goal is to determine the locations where growth occurs. This goal cannot be attained using coordinate system-based approaches, as results will change with the coordinate system adopted and there is no way to know which coordinate system is preferred in any given analysis. The approach presented here provides coordinate system-free, valid information about growth patterns.

    Nearly all aspects of EDMA developed in previous chapters can be applied to the study of growth. Consequently, the essential ideas required to use EDMA in the study of growth have already been presented. We reinforce that knowledge by presenting an outline of the method again, this time using growth data and a specific notation for growth analyses. Let us begin with a single form, A, at two points during ontogeny, time 1 and time 2. The morphology of A at the first point in time is referred to as A_1 while the morphology of A at the second point in time is referred to as A_2. Data are collected for three two-dimensional landmarks at each point in time. For example, the coordinates from these three points could be collected as:

$$A_1 = \begin{bmatrix} 0 & 0 \\ 1 & 0 \\ 0 & 1 \end{bmatrix} \qquad A_2 = \begin{bmatrix} 0 & 0 \\ 3 & 0 \\ .75 & 2 \end{bmatrix}$$

    $FM(A_1)$ and $FM(A_2)$ denote the form matrices corresponding to these landmark data. Recall that $FM_{ij}(A_1)$ is the distance between landmarks i and j in object A_1 and that $FM_{ij}(A_2)$ is the corresponding distance in object A_2. Using the landmark coordinate data given above, we obtain the following form matrices:

$$FM(A_1) = \begin{bmatrix} 0 & 1 & 1 \\ 1 & 0 & 1.41 \\ 1 & 1.41 & 0 \end{bmatrix} \quad FM(A_2) = \begin{bmatrix} 0 & 3 & 2.136 \\ 3 & 0 & 3.010 \\ 2.136 & 3.010 & 0 \end{bmatrix}$$

In Chapter 4 we presented three possible ways to express form difference using the form matrices. We have found that relative change is a useful way of describing differences that occur due to growth and therefore present that approach exclusively. The other approaches described in Chapter 4 could be used to study growth, but appropriate precautions may be necessary when interpreting the results.

5.5 Growth measured as relative form difference

As described previously, the relative difference between two forms can be written as a form difference matrix, $FDM(A_2, A_1) = \left[\dfrac{FM_{ij}(A_2)}{FM_{ij}(A_1)} \right]$ where the elements of the matrix correspond to the ratio of like-linear distances and the division is done element-by-element. In this application, the forms being compared are members of a growth series, so we define the growth matrix (GM) as $GM(A_2, A_1) = \left[\dfrac{FM_{ij}(A_2)}{FM_{ij}(A_1)} \right]$, . Note the similarity in notation and organization of the FDM and the GM. Note also that the older (i.e., larger) form occupies the numerator position. We adopt this convention because it is intuitive to think of values greater than 1 corresponding with a form getting larger through time. Adopting this convention means that when a linear distance increases in size with advancing age, the ratio corresponding to growth of that linear distance is greater than 1. This convention will be followed here but does not need to be followed by users.

$$GM(A_2, A_1) = \begin{bmatrix} 0 & 3 & 2.136 \\ 3 & 0 & 2.13 \\ 2.136 & 2.13 & 0 \end{bmatrix}$$

Continuing with the example data sets given previously:

Each entry represents the ratio of like linear distances in $FM(A_1)$ and $FM(A_2)$. Because growth matrices are square-symmetric matrices with zeros along the diagonal, only the above-diagonal elements are needed to describe form difference due to growth. We write the growth matrix as a vector where only the above-diagonal elements are reported:

$$GM(A_2, A_1) = \begin{bmatrix} 3 \\ 2.136 \\ 2.129 \end{bmatrix}$$

What this *GM* tells us is that the distance between landmarks 1 and 2 is three times larger in form *A* at time 2 than it was at time 1. The distance between landmarks 1 and 3 and between landmarks 2 and 3 is just over two times larger at time 2 than at time 1.

5.6 Estimation of growth using EDMA

The example above uses information from a single form at two different points in time. In certain settings (e.g., clinical), growth of an individual may be of interest, but researchers rarely operate under the assumption that analysis of a single individual is a valid representation of growth for all individuals. Common sense and previous research tells us that growth of a single individual cannot be assumed to represent growth for a population. It is at this point that the difference between longitudinal and cross-sectional data becomes apparent. Sudden spurts and long periods of stasis mark growth of the individual. Even though some of these episodes are held generally in common within members of a species (e.g., the adolescent growth spurt in humans), the exact timing and character of these episodes varies from individual to individual. Consequently, any statistical analysis of growth, whether the data are cross-sectional, longitudinal, or mixed longitudinal, will obscure individual patterns and may smooth the growth pattern for the sample. A measure of the variance will provide some indication of individual differences but will not reconstruct individual patterns of growth. If a generalized pattern of growth is the desired outcome of the analysis, the procedure described below is appropriate. If variability in individual growth patterns is the focus of research, it may be more useful to study the growth of individuals separately and then compare the individual patterns. Such comparisons can be accomplished using Growth Difference Matrix Analysis (see following sections).

Since the true mean forms for any population are never available (see Chapter 3), we need to collect data from many individuals and analyze representative growth patterns using mean forms estimated from these data. Typically a sample of observations is available from which we estimate the parameters. Let us denote the estimated form matrix corresponding to the mean of the first age group by $FM(\hat{A}_1)$. Estimation of the form matrix is discussed in Chapter 4. The estimated form matrix corresponding to the mean of the second group is represented by $FM(\hat{A}_2)$. The estimated growth matrix is given by

$$GM(\hat{A}_2, \hat{A}_1) = \left[\frac{FM_{ij}(\hat{A}_2)}{FM_{ij}(\hat{A}_1)} \right]$$

Although it is expected that scientists present a valid statistical test of their analytical work, in the study of growth, it is likely that an older form will be statistically different from a younger form. Both hypothesis testing and the confidence interval approach to the statistical comparison of forms presented in Chapter 4 can be used for growth data. However, we advocate the use of confidence intervals for the study of growth because simple statistical testing of similarity in form for different age groups doesn't provide us with much new information. The important information that can be obtained from the study of growth concerns the discovery of local similarities or differences in form between age groups. Localization of those measures that change the most, and those that change the least over time can be accomplished by using confidence intervals.

Information pertaining to magnitude of change local to landmarks can be obtained by simple inspection of the GM. Magnitude is measured by the ratio reported in a GM as the relative change in a given linear distance. Consequently, we can speak of a linear distance doubling (ratio of 2.0), increasing by 25% (ratio of 1.25), or decreasing by 5% (ratio of .95) during growth. The direction of change for any specific linear distance occurs along the given distance, but by looking at groups of landmarks and their associated distances or the association of a single landmark with all others, overall directions of change of anatomical structures in relation to others can be inferred. The "delete one" landmark approach (Chapter 4) can be used to determine the relative importance of the various loci and directions of change. These procedures were presented in detail in Chapter 4. More information for interpreting directions of change from given data sets using these procedures will be included in the examples at the end of this chapter.

5.7 Statistical analysis of form and shape difference due to growth

As discussed in previous sections, although statistical testing is a standard part of any quantitative analysis of morphology, it should not be considered the definitive answer to questions concerning similarity of growth patterns. If GDM $(A_2,A_1 : B_2,B_1)$ does not equal 1, patterns of differences from 1 should be sought by close examination of the GDM. This process is labor intensive but graphic aids are available (some of which are discussed in Chapter 4; also see Cole and Richtsmeier, 1998). To test for the statistical significance of differences in shape change due to growth we use the statistic G_{obs}:

$$G_{obs} = \frac{\max GDM_{ij}(A_2,A_1):(B_2,B_1)}{\min GDM_{ij}(A_2,A_1):(B_2,B_1)}$$

where max (= maximum) is the ratio with the largest value within the GDM, and min (= minimum) is the ratio with the smallest value. This test statistic is scale invariant thereby eliminating the effects of scaling. Growth patterns are interpreted as being more similar in shape change as G_{obs} approaches 1. A null hypothesis of similarity in shape change due to growth is stated as:

$H_0 : GM_{ij} (A_2, A_1) = cGM_{ij} (B_2, B_1)$

where c is some scaling factor and $c > 0$. This is a one-way hypothesis that states that the growth of sample B is similar to the growth of sample A. If accepted, it does not imply that the growth of sample A is similar to the growth of sample B. This would have to be tested by restating the null hypothesis as:

$H_0 : GM_{ij} (B_2, B_1) = cGM_{ij} (A_2, A_1)$

and rerunning the analysis using the other sample as the base population.

The one-way nature of this hypothesis demands a specific design for the statistical testing of G_{obs}. Our design uses an extension of the bootstrap approach presented in Chapter 4. The statistical comparison of growth patterns from two populations using a one-way hypothesis requires that one of the samples be chosen to represent the base population. The choice of which sample to use as the base population can be made for statistical reasons (e.g., the sample with the largest number of observations), or on biological grounds (e.g., the sample hypothesized as more primitive in evolutionary studies or the unaffected sample in studies of pathological specimens). Although all four data sets are used to calculate G_{obs}, only the base population is used to produce the bootstrapped samples.

In testing G_{obs}, we start with four samples (A_1, A_2, B_1, B_2) of size m, n, o, and p, respectively. A landmark coordinate matrix that is rewritten as a form matrix for analysis represents each individual in these samples. If population A is chosen as the base population, using simple random sampling with replacement, we obtain a sample of size m from A_1 and of size n from A_2. These bootstrapped samples are designated A_1^* and A_2^*. Similarly, we obtain simple random samples with replacement of size o from A_1 and of size p from A_2. These samples are designated B_1^* and B_2^*. Next, we calculate a growth difference matrix for the bootstrapped samples:

$$GDM_{ij}(A_2^*, A_1^* : B_2^*, B_1^*) = \frac{GM_{ij}(A_2^*, A_1^*)}{GM_{ij}(B_2^*, B_1^*)}$$

and write the GDM as a vector whose elements are sorted from minimum value to maximum value. These steps are repeated B times where B is sufficiently large ($100 \leq B \leq 1000$). In matrix form, this collection of vectors formed by the analysis of B bootstrapped samples has K (K - 1)/2 rows and B columns, each column being a sorted (in ascending order) growth difference matrix for a set of four bootstrapped samples. The statistic G for each of these bootstrapped samples is defined as the maximum ratio in the GDM divided by the minimum ratio in that GDM. G is calculated for each column and represents a summary measure of similarity in shape change due to growth for each bootstrapped set of samples. The placement of G_{obs} calculated from the original data in relation to the distribution of the bootstrapped G values provides a means for determining if growth is similar in the two groups being compared. One may also calculate the probability of getting the observed (or more extreme) value of the test statistic, G_{obs}, and report it as a p-value. Our goal is to determine the probability of obtaining the observed (or a larger) maximum-to-minimum ratio value due to underlying biological variability when the growth of the two groups is, in fact, similar. If this probability, the p-value, is small, we claim that the observed value is unlikely under the null hypothesis of similarity in growth and reject the null hypothesis. By the same results, we entertain the alternative hypothesis.

It is obvious from previous discussions that we are playing down the importance of hypothesis testing. While providing these testing procedures in order to offer a more complete set of tools, our ultimate goal is to present methods for examining *how* two growth patterns differ or coincide with one another. Specifically, we want to *localize* these similarities and differences to anatomical complexes. The identification of specific structures enables interpretation of the biological mechanisms that might underlie these differences. This information could ultimately be used to propose explanatory hypotheses.

There are several ways to localize the difference in growth patterns to anatomical positions. The simplest, but most time-consuming way, is to simply study the individual *GMs* for the samples of interest and the *GDM*. In addition, non-parametric confidence intervals can be calculated for growth differences local to each linear distance following methods similar to those developed for studying form and shape difference (Chapter 4).

Since confidence intervals are not one-way in nature, all four samples in a growth analysis are used to create the bootstrapped samples. Continuing with the previous notation using samples A_1, A_2, B_1, B_2 of size m, n, o, and p, respectively, we obtain a sample of size m from A_1 and of size n from A_2 using a random sampling design with replacement. Similarly we obtain simple random samples with replacement of size o from B_1 and of size p from B_2. A *GDM* is calculated from these bootstrapped samples and then sorted according to the landmarks that define the linear distance. This is done B times. A collection of boot-strapped samples sorted in this way has $\dfrac{K(K-1)}{2}$ rows and B columns in matrix format. Each column is a growth difference matrix for a set of bootstrapped samples and each row represents B growth difference matrix ratios for a defined linear distance. To obtain a confidence interval for each linear distance, the ratios in each row are sorted in increasing order. If a 10% confidence interval is sought, then the first 5% (those of minimal value) and the last 5% (those of maximum value) of the total entries in the row are deleted. The minimum and maximum entries remaining in that row constitute the minimum and maximum values for a bootstrap confidence interval for the linear distance under consideration. As in the confidence intervals for the study of form (Chapter 4), the collection of confidence intervals for all linear distances that constitute a growth difference matrix does not represent a confidence band for growth of the form. These represent marginal confidence intervals calculated separately for each linear distance.

Although this analysis was developed specifically for the comparison of growth patterns, it can be used to study related problems in biology. For example, Lague and Jungers (1999) used *GDMA* to study differences in patterns of sexual dimorphism in joint surface morphology among hominoid species. Other problems that require a quantitative method for comparing comparisons may also be studied using the method.

5.8 Growth difference matrix analysis: comparing patterns of growth using growth matrices

"Great as are the differences between the rates of growth in different parts of a complex organism, the marvel is that the ratios between them are so nicely balanced as they are, and so capable of keeping the form of the growing organism all but unchanged for long periods of time, or of slowly changing it in its own harmonious way. There is the nicest possible balance of forces and resistances in every part of the complex body; and when this normal equilibrium is disturbed, then we get abnormal growth, in the shape of tumours and exostoses, and other malformations and deformities of every kind."

D'Arcy Thompson (1992)

Our interest may be less in defining a single growth pattern as in understanding how the growth pattern for one group differs from that of another. For example, an abnormal growth pattern needs to be compared to a normal growth pattern to be understood. The comparison of growth patterns can only be done in a coordinate system-free method of measurement. Our presentation of a coordinate system-free system for the comparison of growth patterns follows methods previously presented by Richtsmeier and Lele (1993).

To compare growth patterns of two populations, four samples are used in analysis. Each sample must contain data for the exact same landmarks. Let us say that we have collected sample data from individuals of population A from two age groups, 1 and 2. We refer to these samples as A_1 and A_2. We have also collected sample data from corresponding age groups in population B. We refer to these samples as B_1 and B_2. In this example, we assume that the populations have been aged correctly, according to comparable aging criteria.

The first step is to calculate the growth pattern for the samples that represent population A. The growth matrix is estimated using two form matrices, one for sample A at time 1 and the other for sample A at time 2:

$$GM(A_2, A_1) = \left[\frac{FM_{ij}(A_2)}{FM_{ij}(A_1)} \right]$$

The growth matrix for samples representing population B is estimated similarly using the estimated form matrices for the appropriately aged samples:

$$GM(B_2, B_1) = \left[\frac{FM_{ij}(B_2)}{FM_{ij}(B_1)} \right]$$

We now have two growth matrices, each representing an estimate of the growth pattern for a population. The comparison of two growth matrices, one for growth of A_1 into A_2 and one for growth of B_1 into B_2, define the differences in relative growth. Growth difference matrix analysis ($GDMA$) enables comparison of these growth patterns on a linear distance-by-linear distance basis. The results of $GDMA$ report how change in a linear distance in one sample compares to the change experienced by the same linear distance in another sample over comparable growth intervals. $GDMA$ does this for all linear distances. $GDMA$ uses the two GMs from the samples under study to estimate a growth difference matrix (GDM):

$$GDM(A_2, A_1 : B_2, B_1) = \frac{GM_{ij}(A_2, A_1)}{GM_{ij}(B_2, B_1)} = \frac{FM_{ij}(A_2)/FM_{ij}A_1)}{FM_{ij}(B_2)/FM_{ij}(B_1)}$$

In this notation, two organisms, or samples of organisms have the same relative growth if and only if the GDM consists exclusively of ratios equal to 1.

5.9 Analysis of example data sets: differences in facial growth

To demonstrate these various techniques for studying growth and differences in growth patterns, we use the data sets introduced in Chapter 1. We provide an analysis of facial growth for a species of New World monkey, *Cebus apella,* and a species of Old World monkey, *Macaca fascicularis.* Individuals from both species were aged according to tooth eruption patterns so that we are comparing equivalent developmental episodes in these two species. The growth episode considered here spans the time elapsed from the earliest developmental group for which we have data (individuals who have any deciduous teeth present but not the first permanent molar) to adulthood (individuals whose permanent dentition is fully erupted including both the maxillary and mandibular third molars). We understand that a growth analysis with only two points along a time continuum is insufficient, but we present these data as an example. In addition, we remind the reader that our data are cross-sectional and not longitudinal.

Growth of the face in these species is studied using six landmarks located on the sagittal plane and right side of the face (Figure 5.3). Facial growth of *Cebus apella* from immature to adult is represented by the growth matrix, GM_{CA}:

	NSL	IDS	PMM	ZMS	MXT	PNS
NSL	0.00000					
IDS	1.34498	.00000				
PMM	1.23602	1.25348	0.00000			
ZMS	1.23616	1.38671	1.43306	0.00000		
MXT	1.22091	1.32611	1.31896	1.29656	0.00000	
PNS	1.19061	1.31221	1.27509	1.25223	1.20850	0.00000

The distance between zygomaxillare superior (ZMS) and premaxillary-maxillary junction (PMM) shows the largest magnitude of change relative to all other linear distances. In addition, the growth matrix shows that a relatively large degree of change has occurred between intradentale superior (IDS) and all other landmarks. The amount of growth occurring between nasale (NSL) and other landmarks (excepting that between NSL and IDS) is relatively small. A very small amount of change occurs local to the distance between posterior nasal spine (PNS) and maxillary tuberosity (MXT), a distance representing the width of the posterior palate.

Facial growth from immature to adult *Macaca fascicularis* is summarized in the growth matrix GM_{MF}:

	NSL	IDS	PMM	ZMS	MXT	PNS
NSL	0.00000					
IDS	1.85141	0.00000				
PMM	1.44763	1.30635	0.00000			
ZMS	1.45993	1.99816	2.13650	0.00000		
MXT	1.45347	1.71529	1.81614	1.51753	0.00000	
PNS	1.32945	1.66516	1.64113	1.26350	1.48338	0.00000

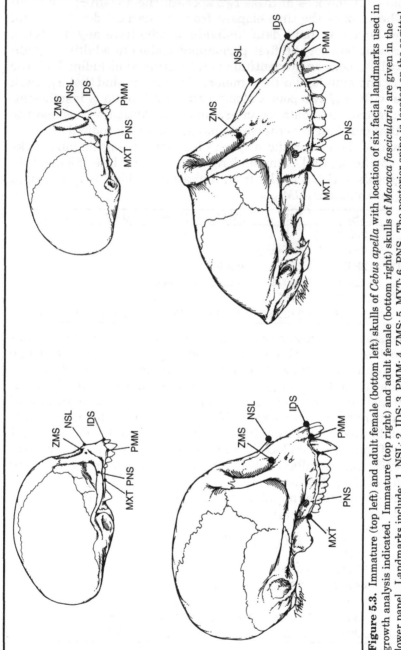

Figure 5.3. Immature (top left) and adult female (bottom left) skulls of *Cebus apella* with location of six facial landmarks used in growth analysis indicated. Immature (top right) and adult female (bottom right) skulls of *Macaca fascicularis* are given in the lower panel. Landmarks include: 1, NSL; 2, IDS; 3, PMM; 4, ZMS; 5, MXT; 6, PNS. The posterior spine is located on the sagittal plane and cannot be seen from this view. Its location in this drawing is therefore approximate. Though both immature specimens are from the same developmental age group, the immature *M. fascicularis* skull is of a younger age than the immature *C. apella* skull, which accounts for the size difference in these immature specimens.

Facial growth in *Macaca fascicularis* is characterized by large magnitudes for those distances from ZMS to IDS and to PMM. These linear distances measure the degree of prognathism in the *Macaca fascicularis* face, or the degree to which the lower face projects from the orbits. The distance between NSL and IDS changes substantially during growth. The linear distances indicate a principal extension of the anterior portion of the maxillary alveolus in relation to the midface asa result of growth from early postnatal life to adulthood. The distance between the midface and posterior palate (measured from ZMS to PNS) experiences the smallest magnitude of change during this time interval.

Facial growth in *Macaca fascicularis* is of a greater magnitude than the facial growth occuring in *Cebus apella*. This is not surprising given that *Macaca fascicularis* adults are considerably larger than *Cebus apella* adults. Estimates of adult body weight for wild caught *Cebus apella* males range from 2646 to 3300 g (Napier and Napier, 1967; Leutenegger and Cheverud, 1982; Harvey and Martin, et al., 1987) while *Macaca fascicularis* adult males achieve an adult body weight of approximately 4930 g (Fleagle, 1988).

Simply looking at similar entries in the two growth matrices provides a comparison of growth patterns for these two species. However, comparison of these two relatively small matrices is tedious and this task becomes increasingly difficult as the number of landmarks increases. Moreover, we would like to be able to compare patterns within these matrices, rather than just single elements. A growth difference matrix that compares growth for these two species, $GDM_{CA\text{-}MF}$, is given below.

	NSL	IDS	PMM	ZMS	MXT	PNS
NSL	0.00000					
IDS	0.72646	0.00000				
PMM	0.85382	0.95952	0.00000			
ZMS	0.84672	0.69399	0.67075	0.00000		
MXT	0.84000	0.77311	0.72624	0.85439	0.00000	
PNS	0.89557	0.78804	0.77696	0.99109	0.81469	0.00000

When the elements of this growth difference matrix are arranged in ascending order, we obtain the following vector:

Linear Distance	Growth Difference Ratio
PMM-ZMS	0.671
IDS-ZMS	0.694
PMM-MXT	0.726
NSL-IDS	0.726
IDS-MXT	0.773
PMM-PNS	0.777
IDS-PNS	0.788
MXT-PNS	0.815
NSL- MXT	0.840
NSL-ZMS	0.847
NSL-PMM	0.854
ZMS-MXT	0.854
NSL-PNS	0.896
IDS-PMM	0.960
ZMS-PNS	0.991

Since all elements of the growth difference matrix are less than 1, we know that the species represented by the numerator, *Cebus apella*, experiences a smaller magnitude of change during growth in every facial dimension as compared to *Macaca fascicularis*. We also observe that there is a great degree of variability among the elements of the growth difference matrix indicating that the difference in facial growth between these two species cannot be adequately defined as a difference in scale.

Inspection of the growth difference matrix indicates that the magnitude of growth in the two species is similar for the distance between ZMS and PNS (the ratio approaches 1). In both species, this is a linear distance that experienced a relatively small degree of change during growth. Growth from IDS to PMM was moderately low in both species and the growth difference matrix shows that the magnitude of growth is similar in the two species.

Looking at the minimal end of the matrix, we note two linear distances (ZMS to PMM and ZMS to IDS) that experienced a relatively large degree of growth for each species. The growth difference matrix underscores the difference in the magnitudes of growth experienced by each species local to these distances. The distance between PMM and MXT measures the length of the maxillary alveolus posterior to the incisors. Increase in length of the alveolus due to growth is far greater in *Macaca fascicularis* as indicated by the growth difference matrix.

In certain instances, an element-by-element examination of each growth matrix alongside the growth difference matrix is the best way to digest the analytical output. Statistical tools can also help in assimilating this information as shown below.

5.9.1 Statistical testing of similarity in growth pattern

Hypothesis testing for similarity in growth pattern

The range of the elements of the growth difference matrix just presented (0.671 - 0.991) indicates that the difference in facial growth in these two species is not simply a matter of scale. Our previous discussion suggests that there are significant differences in facial growth patterns of these two species. If statistical testing for difference in overall growth pattern is required or desired by the investigator, G_{obs} can be calculated following methods outlined above and detailed in Chapter 4 for the two-sample case.

In this example, the Bootstrap reference sample is *Macaca fascicularis* (the denominator). The histogram below provides the distribution of 200 bootstrapped G statistics (each "*" represents the value of a single bootstrapped G statistic) as well as the placement of G_{obs} (signified by the letter G in the figure) within this distribution. The probability is given as 0.204 and G_{obs} clearly falls within the central tendency of the distribution. This means that we cannot reject the null hypothesis that facial growth of *Cebus apella* is similar to the facial growth of *Macaca fascicularis*. Here is a clear example where biologically interpretable differences have been revealed using the growth matrices and growth difference matrix analysis, but statistical hypothesis testing suggests that the overall difference in growth is not significant.

We saw earlier that there were similarities in the pattern of facial growth of these two species. Those linear distances that changed the most and that changed the least were similar in the two species. Even though the magnitudes of the changes differed greatly, the overall *pattern* of growth is similar between the two species. From this perspective the results of the G_{obs} testing is not unexpected.

G_{obs} statistic (max/min): 1.478
p-value: 0.204

```
1.047    |******
1.152    |*********************************************************
1.257    |***********************************************
1.362    |****************************************
1.468    |G************************
1.573    |**************
1.678    |*******
1.783    |***
1.888    |*
1.994    |*****
2.099    |
2.204    |*
2.309    |*
2.415    |*
2.520    |
2.625    |
2.730    |*
2.836    |
2.941    |*
3.046    |*
```

Confidence intervals for GDMA

Although we have some idea of the magnitude of differences in growth pattern by inspecting the growth difference matrix, this cannot provide information regarding which of these growth magnitudes is statistically different between the samples. Large within-sample variability can underlie estimates of the mean and this may produce unexpected statistical results. As explained previously, confidence intervals can be calculated for the comparison of growth patterns for each linear distance. Confidence interval testing for differences in facial growth in *Cebus apella* and *Macaca fascicularis* (shown below) is based on 200 non-parametric Bootstrap resamples. The chosen alpha level is 0.10. If growth for a specific linear distances is significantly different in the two samples, that linear distance is marked with a "**"."

Table 5.1 Confidence Intervals for Growth Differences $\alpha = 0.10$

		Low	Estimate	High	
NSL	IDS	0.644	0.726	0.805	**
NSL	PMM	0.794	0.854	0.915	**
IDS	PMM	0.867	0.960	1.041	
NSL	ZMS	0.791	0.847	0.907	**
IDS	ZMS	0.606	0.694	0.774	**
PMM	ZMS	0.573	0.671	0.772	**
NSL	MXT	0.773	0.840	0.903	**
IDS	MXT	0.701	0.773	0.843	**
PMM	MXT	0.656	0.726	0.792	**
ZMS	MXT	0.789	0.854	0.913	**
NSL	PNS	0.837	0.896	0.950	**
IDS	PNS	0.717	0.788	0.849	**
PMM	PNS	0.718	0.777	0.835	**
ZMS	PNS	0.938	0.991	1.046	
MXT	PNS	0.729	0.815	0.904	**

These results show that all but two linear distances experience a significantly greater degree of growth in the *Macaca fascicularis* sample. Taken together, we find that we cannot reject the null hypothesis that the overall pattern of growth is similar between the two groups, but the magnitude of growth measured local to specific linear distances is significantly different between the two species.

5.9.2 A method for graphically displaying differences in growth trajectories

The *GDM* compares the change in each linear distance due to growth in one sample to the change experienced in those same linear distances during growth in the other sample. As detailed above, elements of the *GDM* can be sorted from minimum to maximum, providing the ranking of those landmarks or regions that contribute most strongly to the differences in growth between the two samples under consideration. Graphs like those shown in Figure 4.11 can be used to identify landmarks or linear distances that show patterns of growth that stand out in relation to others. Once a specific set of linear distances has been identified as "interesting," the relative changes in these linear distances can be compared graphically as trajectories.

To maintain simplicity and clarity, we limit ourselves to graphing triplets of linear distances, each axis tracking the change in a single

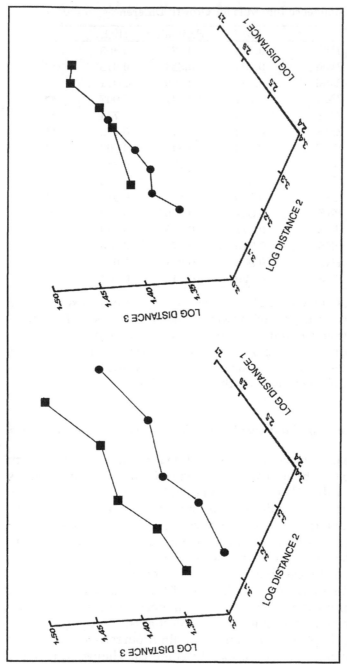

Figure 5.4. A method to graphically compare growth of three linear distances (expressed in natural logarithms). Graph on left shows ideal expectation of parallel lines when growth is similar in two organisms or two populations. Graph on right shows longitudinal growth data for three linear distances measured on the cranial base of two New Zealand white rabbits at 5, 7, 9, 11 and 13 weeks of age. The two rabbits were raised together on a similar diet. Each point on the graph represents the measure of three linear distances at a specific age. Data from rabbit 1 are plotted as circles; data from rabbit 2 are plotted as squares. Data appear in chronological order from lower left to upper right. Although growth of these two rabbits appears similar, these graphs are not invariant to rotation and should be viewed from many perspectives in order to determine the similarity of the growth patterns displayed.

linear distance. In such a graph, equivalent patterns of relative growth in two groups are represented by a constant in the *GDM*, and the two growth patterns are graphically represented as parallel vectors when linear distances are plotted in natural log space (see Figure 5.4a). If growth patterns differ between the two samples being compared, elements of the *GDM* are not uniform and vectors representing the growth patterns are not parallel.

5.10 Producing hypothetical morphologies from forms and growth patterns

An additional application of the methods provided in this chapter concerns the production of hypothetical forms using growth patterns quantified by EDMA. There are many instances where the creation of hypothetical forms may be useful in research. Suppose that there is a disease that results in dysmorphology. The mutation that causes this disease has been identified, but it is unclear whether this mutation causes a primary insult that continually affects systemic or localized growth patterns of affected individuals or if the mutation results in a prenatal developmental insult that causes a dysmorphology which is then maintained by a normal growth pattern. Morphological data could be used to explore these possibilities. For example, a growth matrix calculated from a sample of normal individuals can be applied to coordinate data representing the morphology of an affected individual. In a sense, we are "growing" an affected individual according to a growth pattern derived from a sample of unaffected individuals. The resulting form represents the morphology of an affected individual who has followed a "normal" growth pattern. The hypothetical form produced using this method can be statistically compared to age matched affected and unaffected individuals to determine the difference in form. Similarly, the hypothetical form and samples of normal and affected individuals can be analyzed using clustering or classification techniques (see Chapter 6) to place the hypothetical form into the group that it matches most closely. These analyses provide information regarding the role of growth patterns in producing dysmorphology.

   Another example of the use of this methodology for the production of hypothetical forms concerns the testing of hypotheses about the role of heterochrony in evolution. To offer a simplified sample, let us say that species A has been described as evolving from species B through the process of hypermorphosis ("going beyond" the morphology of species B; see Gould, 1977). We can continue to grow the adult form of species B by applying the complete growth pattern or a specific portion of the growth pattern for species B to the species B adult morphology, producing a hypothetical form that we refer to as B´. B´ can be compared to the adult form of species A. This comparison will determine whether an extension of the growth pattern of species B can produce a morphology consistent with that of species A.

To provide an example of how EDMA can be used to produce hypothetical forms, we take the growth matrix obtained for *Macaca fascicularis* facial growth and apply it to the landmark data representing the face of the youngest *Cebus apella*. The form matrix for the starting form (*Cebus apella* age 1) is:

Form matrix for the starting form FM_{Ca_1}

0.00000					
1.28355	0.00000				
1.36297	0.77527	0.00000			
1.55937	1.97504	1.34787	0.00000		
2.67187	2.62567	2.23906	1.86376	0.00000	
2.24435	2.33153	2.16739	1.96148	0.83985	0.00000

The "transformation matrix" used in this example is the growth matrix obtained by comparing the young to the old *Macaca fascicularis* face (GM_{MF}).

Transformation Matrix GM_{MF}

0.00000					
1.85100	0.00000				
1.44800	1.30600	0.00000			
1.46000	1.99800	2.13600	0.00000		
1.45300	1.71500	1.81600	1.51800	0.00000	
1.32900	1.66500	1.64100	1.26300	1.48300	0.00000

The matrix obtained by multiplying the chosen form matrix FM_{Ca_1} by the transformation matrix (GM_{MF}) represents the hypothetical form matrix.

Hypothetical Form Matrix

0.00000					
2.37584	0.00000				
1.97359	1.01250	0.00000			
2.27668	3.94614	2.87906	0.00000		
3.88223	4.50303	4.06614	2.82918	0.00000	
2.98274	3.88200	3.55668	2.47735	1.24550	0.00000

If we want to plot the relative landmark locations of the hypothetical form in three-dimensional space, we need the coordinates for the landmark locations of the hypothetical form. To re-write the hypothetical form matrix as a landmark coordinate matrix, the form matrix is subjected to spectral decomposition (see also Chapter 3, Part 2). To do this we first square the Euclidean distance matrix corresponding to the hypothetical form matrix. Following the notation given in Part 2 of Chapter 3, this is written as:

$$HFM(M) = \begin{bmatrix} 0 & \delta_{12}^2 & \cdots & \delta_{1K}^2 \\ \delta_{12}^2 & 0 & \cdots & \vdots \\ \vdots & \vdots & \vdots & \vdots \\ \delta_{1K}^2 & \delta_{2K}^2 & \cdots & 0 \end{bmatrix}$$

The squared Euclidean distance matrix corresponding to the hypothetical form in our example is given as:

0.00000					
5.6446	0.00000				
3.89505	1.0251	0.00000			
5.18327	15.5720	8.2889	0.00000		
15.0717	20.2773	16.5335	8.00426	0.00000	
8.89674	15.0699	12.64997	6.13726	1.55127	0.00000

Next, we need to calculate the corresponding centered inner product-matrix using

$$B(M) = -\frac{1}{2} H[EM(M)]H^T$$

where $H = \begin{bmatrix} 1-\dfrac{1}{K} & -\dfrac{1}{K} & \cdots & -\dfrac{1}{K} \\ -\dfrac{1}{K} & 1-\dfrac{1}{K} & \cdots & \vdots \\ \vdots & \vdots & 1-\dfrac{1}{K} & -\dfrac{1}{K} \\ -\dfrac{1}{K} & -\dfrac{1}{K} & \cdots & 1-\dfrac{1}{K} \end{bmatrix}$ is a $K \times K$ symmetric matrix.

The eigenvalues $(\lambda_1 > \lambda_2 > \lambda_3 > ... \lambda_K)$ and eigenvectors of $B(M)$ are calculated next. The eigenvalues are arranged in decreasing order by λ_1, $\lambda_2, \lambda_3, ..., \lambda_K$, and the corresponding eigenvectors are given as $h_1, h_2, ...,$ h_n. Note that h_1, h_2, and h_3 are K x 1 vectors, and $\lambda 1$, $\lambda 2$, and $\lambda 3$ are positive real numbers provided the hypothetical form is a three-dimensional object. If the original data are from two-dimensional objects, the estimate of the coordinates of the hypothetical form matrix is given by $M = [\sqrt{\lambda_1}h_1, \sqrt{\lambda_2}h_2]$. If the original data are from three-dimensional objects the estimate of the coordinates of the hypothetical form matrix is given by $M = [\sqrt{\lambda_1}h_1, \sqrt{\lambda_2}h_2, \sqrt{\lambda_3}h_3]$.

The landmark coordinate locations for the hypothetical form from our example are given below.

Coordinates of the Hypothetical Form

0.93655	-0.89435	0.88308
2.16769	1.01905	0.03706
1.66302	0.10050	-0.66443
-0.77885	-1.54961	-0.51053
-2.33166	0.81673	-0.32909
-1.65674	0.50768	0.58390

These coordinate data for the hypothetical form can be translated and rotated so that the form appears in a biologically sensible orientation when viewed. Coordinate locations obtained in this way can be used to graphically represent the hypothetical form, but we stress that the estimator is a representation of the hypothetical form only up to rotation, reflection, and translation. Keep in mind that to obtain a graphical representation that is biologically meaningful, the coordinate matrix may need to be reflected. This can be done by multiplying one or more of the axes by -1.

It is important to remember that transformation matrices can be incompatible with form matrices. When this occurs, the hypothetical form produced is of a different dimension than the starting form. To avoid such errors, it is important to check for the dimensionality of the hypothetical form. Checking the dimensionality of a form is a necessary step in the production of hypothetical forms (Richtsmeier and Lele, 1993). The first thing that should be done is to check that the eigenvalues are all positive. There should be no large negative eigenvalues. If there are small, negative eigenvalues, the researcher must choose whether or not to ignore them. If there are large negative eigenvalues, then the growth matrix and the form matrix are incompatible and the geometry of the hypothetical form cannot be resolved. If the majority of the eigenvalues are non-negative, then the dimension of the hypothetical form must be determined.

To determine the dimensionality of a hypothetical form, we suggest adding the first D eigenvalues (where $D=$ the number of dimensions of the starting form) to see if these eigenvalues constitute a given percentage of the sum of the K eigenvalues. We suggest the following criterion be used in determining the dimensionality of a hypothetical form. Let

Choose that value of D that makes P_D larger than 0.95. The 95% cutoff

$$P_D = \sum_{i=1}^{D} \lambda_i / \sum_{i=1}^{K} \lambda_i$$

is arbitrary, but sensible. A more or less restricted dimensionality check may be required depending upon the purpose of the experiment. If we expect a two-dimensional hypothetical form, then the sum of the first two eigenvalues should exceed 95% of the sum of all eigenvalues. If we expect a three-dimensional hypothetical form, then the sum of the first three eigenvalues (but not the first two) should exceed 95% of the sum of all eigenvalues. The first 6 eigenvalues and their associated percentages for our hypothetical form are given below. The remaining eigenvalues are 0.0.

Eigenvalues

1.	17.12972	70.35886%
2.	5.17452	21.25391%
3.	1.93256	7.93784%
4.	0.10941	0.44939%
5.	0.00000	0.00000%
6.	0.00000	0.00000%

In our example, the sum of the first D eigenvalues (but not the first D-1 eigenvalues) exceeds the check value (95%). Reconstruction of the landmark coordinates for the hypothetical form is possible.

5.11 Summary

In this chapter, we demonstrated the way in which EDMA can be applied in the study of growth patterns. We described the types of data that may be available to the researcher studying growth and suggested ways to categorize individuals on the basis of life history variables in the absence of chronological ages. We defined the growth matrix, GM, and the growth difference matrix, GDM, and provided a method for comparing patterns of growth. Statistical tests and graphical techniques for comparing growth patterns were also presented. Finally, we provided a method for the production of hypothetical forms from combining a growth matrix and a form matrix.

CHAPTER 6

Classification and Clustering Applications

Traditionally quantification of the morphology of biological organisms has played an important role in Numerical Taxonomy (e.g., Sneath and Sokal, 1973, Reyment et al., 1984). A chief objective of many studies in Numerical Taxonomy is to find groups in the data such that the organisms within a group are more similar to other members of the group than they are to the members of alternate groups. A related but distinct issue of substantial interest to the clinical sciences is the classification of new patients into already defined groups. This practice is clearly relevant to the field of medical diagnostics, but it is also useful in evolutionary biology where a scientist may want to assign a newly found individual to a group; e.g., a clade, a species, a family. In this chapter, we discuss the ways in which the invariant approach to the quantitative analysis of forms can be applied towards the problem of forming groups (clustering) and the assignment of individuals to known groups (classification). In standard statistical terminology, the classification approach is used when the groups are *known* and the goal is to determine the group membership of a new individual. The goal of the clustering approach is to find the groups in the data. There exist a vast array of statistical procedures that can be applied to attain either goal. We start with the problem of classification, in part, because it provides a relatively clean statistical answer. We then turn to clustering, although with less enthusiasm due to the fact that clustering is inherently subjective. This is because the results of a clustering procedure can depend more on the method than on the signal in the data. In this chapter, we propose algorithms for classification and clustering of individuals represented by landmark coordinate data.

The use of these algorithms is illustrated using datasets involving craniosynostosis patients.

6.1 Classification analysis

The activity of allocating individuals to one of a set of existing groups is usually referred to as classification, assignment or discrimination. The term classification has two complementary meanings in the statistical literature (Gower, 1998). First, classification involves assigning an observation to one of a set of previously known classes. Secondly, classification is concerned with the construction and description of the classes themselves. In this book, we define classification in the first sense; i.e., assigning a new observation to one of the known classes. Gower (1998) suggests that this is better described as discrimination and we agree that the theory of discriminant analysis plays an important role in this activity. Here, a discrimination or classification problem concerns the identification of the sample of interest as belonging to one group within a set of known groups. Using a combination of features that are held in common by the individuals in the same group, an individual is identified as belonging or not belonging to the group. The features may be based solely on landmark data or additional information; e.g., life history information, ecological variables, phylogenetic information. In this book, we restrict ourselves to the exclusive use of landmark data. However, modification of the classification procedure to accommodate other variables is straightforward and may be particularly valuable in specific research situations.

Classification involves specification of a rule that determines the class to which a new observation is assigned. There are two kinds of classification rules: "probabilistic" and "non-probabilistic." In a research situation of living organisms, classification based on gender (which is observable from the organism) is non-probabilistic because the membership of a new individual to the class 'male' or 'female' can be done with *certainty*. On the other hand, probabilistic classification involves the assignment of specimens into groups with the possibility for misclassification. For example, classification of an organism into the groups male and female is usually probabilistic when the specimen has been skeletonized (or fossilized), and observable features do not include genitalia; i.e., there is uncertainty inherent to the classification. The main difference between a non-probabilistic and probabilistic classification, then, is that a non-probabilistic classification rule does not involve the possibility of misclassification of the new observation into the 'wrong' class, whereas the probabilistic classification includes the possibility of misclassification. This feature makes probabilistic classification more general, less certain, and more interesting.

Let us consider the probabilistic classification approach for landmark coordinate data. Suppose we have C different classes. The classes might correspond, for example, to different classes (families, clades, species, etc.) of Early Eocene Notharctinae, each individual represented by a fossilized first mandibular molar. Suppose further that for each class a mean form and variance exists representing the first molar. Let us assume that the variance can be written in the Kronecker product form: $V = \Sigma_K \otimes I_D$. Let $(M_i, \Sigma_{K,i})$ denote the mean and variance of the i-th class. Now, suppose we discover an additional first molar in a museum drawer and suspect that the individual represented by this tooth might belong to one of the known classes. In order to classify this tooth, we collect landmark coordinate data from surface features. The question is: to which of the C classes does this individual most likely belong? Notice several important features of this situation. First, we assume that the individual has to come from one of the C classes and no other class. Second, we assume that the mean and the variance for each class are completely known *a priori*. The first assumption may be justified based on the knowledge and expertise of the scientist. The second assumption may be justified based on the estimated means and variances of the previously classified fossil samples whose phylogenetic relationships and class membership are confirmed. However, in practice one or both of these assumptions may be violated.

6.2 Methods of classification

1. **Likelihood-based classification**
   Suppose we accept the two assumptions as reasonable. Suppose further that there are two known classes. Now suppose, under the usual Gaussian perturbation model, we calculate the probability that the new individual belongs to the first class. Let us say, for example, this probability is 0.3. Similarly we calculate the probability that the new individual belongs to the second class and find this probability to be 0.8. Naturally we would assign that individual to the category that has the highest probability, in this case we would

classify the individual to class 2. This type of classification rule corresponds to what is called 'likelihood classification'.

Unfortunately, calculation of the likelihood classification is computationally difficult when landmark coordinate data are used because of the presence of the nuisance parameters of translation, rotation and reflection. The difficulty in computing the likelihood rule is explained in Chapter 3, Part 2.

2. Dissimilarity measures-based classification

Fortunately, likelihood method is not the only approach to classification. Another classification tool uses a dissimilarity measure. In this approach, the first step is to calculate a distance, or dissimilarity measure between the observation of interest and each of the defined classes. This distance, represented by a single number, quantifies the resemblance (or lack of resemblance) between the new observation and the class under consideration. We classify the new observation as belonging to the class that it resembles most closely according to the chosen measure.

The choice of the measure, be it a dissimilarity metric or a distance, is subjective. For each distance measure, there exists a corresponding classification rule. As we noted earlier, every probabilistic classification rule has the potential to lead to a wrong decision. Given two different classification rules, we choose that rule that on average leads to a smaller number of wrong decisions. Unfortunately, in practice, no single classification rule turns out to be uniformly the best, in the sense that it has the smallest error rate under *every possible* situation. It is unfortunate, but true, that some classification rules are good for some situations whereas others work well under different conditions. This feature makes classification more an art than science. There seldom is a single classification rule that is best under every situation. It is futile to argue for one classification rule over others based on a few, simple, specific situations.

There will certainly be some situations where an alternate classification rule works better than the rule we are proposing here. Presentation of situations where one rule provides superior outcomes does not prove the superiority of a given approach. Instead this demonstrates that one classification rule should not be used in every possible instance, and that no particular rule should be discarded universally.

6.3 Dissimilarity measures for landmark coordinate data

In the following paragraphs, we present a few dissimilarity measures that are intuitive, simple to calculate, and useful for classification based on landmark coordinate data. An example of their use in classification is illustrated in the following section.

Let X denote the landmark coordinate matrix for the new observation. In this description, let M denote the true mean form for one of the classes and Σ_K denote the true variance. Recall however from the discussion in Chapter 3, that these values are non-estimable. Only the centered and rotated version of the mean form, M^c, the form matrix, $FM(M)$, and the singular version of Σ_K are estimable. Next we provide various dissimilarity measures for determining the classification of X into one of several known classes. Notice that in all the measures described below only these estimable parameters are used. These dissimilarity measures can therefore be computed in practice.

1. **Unweighted Procrustes dissimilarity measure:** This is a Procrustes distance between the centered landmark coordinate matrix, $X^c = HX$, and the centered mean form matrix, $M^c = HM$, where H is the centering matrix defined in Chapter 3. The distance is given by: $D_P(X^c, M^c) = tr(X^{c^T} X^c) + tr(M^{c^T} M^c) - 2tr(X^c M^{c^T} M^c X^{c^T})^{1/2}$.

   The last component of this expression corresponds to the square root of a matrix. This dissimilarity measure is simple to calculate, however the value of the measure depends on the specific centering adopted during analysis and does not take correlations between landmarks into consideration.

2. **Dissimilarity measure based on the Euclidean Distance Matrix representation using arithmetic differences:** Suppose we calculate the form matrices corresponding to the observation X and the mean form matrix M. A dissimilarity

measure based on the arithmetic differences between the two matrices, D_A, is given by:

$$D_A(X,M) = \sum_{i=1}^{K-1} \sum_{j=i+1}^{K} (FM_{ij}(X) - FM_{ij}(M))^2$$

This measure is simple to calculate and does not depend on any centering protocol. However, it does not take into account the covariance between the landmarks.

3. **Dissimilarity measure based on the Euclidean Distance Matrix representation using relative differences:** A dissimilarity measure based on the relative differences between matrices representing the mean form and an individual is given by:

$$D_R(X,M) = \sum_{i=1}^{K-1} \sum_{j=i+1}^{K} \left(\frac{FM_{ij}(X)}{FM_{ij}(M)} - 1 \right)^2$$

This measure is also simple to calculate but weighs the differences differently; e.g., the smaller distances receive more weight in the calculation. This may or may not be desirable depending upon the research problem.

4. **Dissimilarity measure based on logarithmic differences:** The following dissimilarity measure is based on the differences between the logarithms of the distances:

$$D_{LR}(X,M) = \sum_{i=1}^{K-1} \sum_{j=i+1}^{K} (\log FM_{ij}(X) - \log FM_{ij}(M))^2$$

5. **Dissimilarity measure based on the Euclidean Distance Matrix representation of relative differences in growth pattern:** We have found that there are situations in which a growth pattern can be useful in distinguishing between groups (Richtsmeier and Lele, 1993; Richtsmeier et al. 1993a, 1993b). There are situations in which organisms achieve similar adult forms through differing growth patterns. In this case a dissimilarity measure based on the growth matrices (and the differences between them) can be used. Growth data are required for each individual in question and for each of the defined groups. The dissimilarity

measured based on differences between growth patterns is defined as:

$$D_G(X,M) = \sum_{i=1}^{K-1} \sum_{j=i+1}^{K} \left(\frac{GM_{ij}(X_1, X_2)}{GM_{ij}(M_1, M_2)} - 1 \right)^2$$

The dissimilarity measures provided above are only a sample of possible dissimilarity measures that may be used in classification.

Description of the classification rule

All of the dissimilarity measures given above can be used in devising a classification rule. The steps of the procedure are the same regardless of which measure is used. Suppose, for example, we choose the Procrustes distance to use as the dissimilarity measure. Then, the steps are as follows:

> STEP 1: Collect samples from each of the classes that are of interest. Estimate the mean form matrix and the variance-covariance matrix for each sample using these data. Let $\hat{M}_1$ and $(\hat{\Sigma}_K)$ denote the estimated mean form and the variance covariance matrix for the first sample. Do this for the samples representing each class.

> STEP 2: Given a new individual represented by a landmark coordinate matrix, calculate the Procrustes distance between the individual and each of the classes.

> STEP 3: Classify the new individual into that class that it resembles the most; i.e., the smallest Procrustes distance.

In the next section, we compare the performance of the dissimilarity measures described earlier using data from individuals diagnosed with differing craniofacial malformations.

6.4 A classification example

As discussed in Chapter 1, the developing neurocranium is made up of a number of roughly shell-shaped bony plates that align with one another at joints or articulations called sutures. The typical shape of the calvarium of children diagnosed with premature closure of the sagittal suture is long and narrow relative to a normal skull (Figure 1.4b). Additional sutures are present on the calvarium (Figure 1.4a) and premature closure of any of the other sutures results in abnormally shaped skulls. When the metopic suture is closed prematurely, the frontal bone (forehead) becomes pointed and the skull appears triangular in shape when viewed from above. When either the right or left coronal suture is closed prematurely, the skull takes on an asymmetric shape when viewed from the top with the forehead and orbit compressed on the synostosed side. There can also be flattening of the posterior aspect of the same side and bulging of the neurocranium on the posterior aspect of the unaffected side (see Figure 7.2).

Asymmetric deformation of the developing skull can occur even when all sutures remain open. Typically, these skull deformations mimic the appearance of either premature closure of one coronal suture or one lambdoid suture. Crowding in the womb or repeated placement of the infant in a particular sleeping position can result in flattening of one side of the back of the skull. This deformity is referred to as plagiocephaly (literally, twisted skull) and does not involve craniosynostosis (i.e., all sutures remain open). Data sets for this example include landmark coordinate data collected from samples of individuals diagnosed with posterior plagiocephaly, meaning that the flattened part of their skull was evident on the more posterior aspect, unicoronal synostosis (where the coronal suture closes prematurely), metopic synostosis (where the metopic suture closes prematurely), and sagittal synostosis (premature closure of the sagittal suture) (see Chapter 1). Both unicoronal synostosis and posterior plagiocephaly can occur on either the right or left side of the skull, and our sample contains children with both right and left unicoronal synostosis and right and left posterior plagiocephaly. For the cross validation study presented below, we reflected all right-sided unicoronal and posterior plagiocephaly cases such that all cases were left-sided for analysis.

Most children that present with obviously misshapen skulls are referred to a specialist for consultation. Often a computed tomography scan is necessary to determine whether or not a suture is truly closed and to provide information for clinical correction of the deformity (either by surgery or by a device). Landmark data were collected from computed tomography images of children diagnosed with sagittal synostosis ($N=25$), metopic synostosis ($N=11$), unicoronal synostosis ($N=4$), and posterior plagiocephaly ($N=9$). The landmarks were identified on the cranial base (that part of the skull underlying and surrounding the inferior surface of the brain), a portion of the skull that can be involved (but is not necessarily involved) in the various deformities considered.

Here, we report a cross-validation study of the misclassification rates for the first four measures of dissimilarity described above. Misclassification rates represent the percent of cases misclassified after the application of a specific dissimilarity measure. Misclassification rates indicate the number of times an individual gets classified into an incorrect class. If this rate is high, the dissimilarity measure used is a relatively poor metric for classification, or the data do not provide adequate information for classifying individuals into the proper class. In this experiment, we utilized four different measures of dissimilarity in analysis of the same data set and compared the results. The steps are provided below with the Unweighted Procrustes distance given as the dissimilarity measure. The same steps were followed using each of the four dissimilarity measures. The procedure is as follows:

> STEP 1: Pick one of the categories, say sagittal synostosis. Randomly remove one of the individuals from that sample and call it X.
>
> STEP 2: Estimate the mean form for each of the dysmorphology categories.
>
> STEP 3: Compute the dissimilarity measure between the landmark coordinate matrix for X (the individual which was removed in Step 1) and the estimated mean forms for each of the dysmorphology categories. For the first group, sagittal synostosis, calculate
>
> $$D_P(X^C, M_1^C) = tr(X^{C^T} X^C) + tr(M_1^{C^T} M_1^C)$$
>
> $$-2tr(X^C M_1^{C^T} M_1^C X^{C^T})^{1/2}$$
>
> where M_1 is the estimated mean form for the first group (sagittal synostosis patients). Similarly, compute the dissimilarity between individual X and the estimated mean forms for the remaining groups: unicoronal synostosis (M_2), metopic synostosis (M_3), posterior plagiocephaly (M_4):
>
> $$D_P(X^C, M_2^C), D_P(X^C, M_3^C), D_P(X^C, M_4^C)$$
>
> STEP 4: Classify the individual into the dysmorphology category for which the dissimilarity is the smallest. Check for

correctness of classification and record the outcome (classified correctly or not).

STEP 5: Go back to Step 1, replace the individual that was removed in this step in the previous iteration, remove a different individual and repeat steps 2 to 4. Continue until each individual has been removed once and classified.

This method of studying the error rates of a particular procedure is a cross-validation procedure (Efron and Tibshirani, 1991). We repeated the cross-validation procedure for other dissimilarity measures. Table 6.1 provides the number of times an individual was classified incorrectly. This table suggests that different dissimilarity measures have about the same misclassification rates and that landmark coordinate data from the cranial base may not be sufficient for diagnostic classification of certain craniofacial phenotypes. Interestingly, the unicoronal synostosis class, which is characterized by marked asymmetry of the calvarium and premature closure of a unilateral suture, has the lowest misclassification rate. Posterior plagiocephaly is also identified by asymmetry of the calvarium, but *not* by premature closure of a unilateral suture, and it shows a higher misclassification rate. Both sagittal and metopic craniosynostosis involve fusion of a midline suture, display cranial symmetry, and show relatively high percentages of misclassification.

Table 6.1 Cross-validation based misclassification rates for various dissimilarity measures and four known classes. The smaller percentage indicates a more correct classification. All four dissimilarity measures seem to be equally effective in classifying individuals in different categories of craniofacial anomalies.

	Unicoronal Synotosis $n = 4$	Sagittal Synotosis $n = 25$	Posterior Plagiocephaly $n = 9$	Metopic Synotosis $n = 11$
Procrustes	0%	40%	22%	18%
Arithmetic form difference	0%	36%	33%	18%
Relative form difference	0%	36%	33%	18%
Logarithmic form difference	0%	32%	33%	27%

6.5 Cluster analysis

Cluster analysis is concerned with investigating a set of data to discover whether or not relatively distinct groups of observations can be identified within the data set. The groups are NOT known *a priori*, and this distinguishes cluster analysis from the activity of allocating individuals to one of a set of existing groups or classification as described above. Uncovering the group structure (if any) of a set of data is clearly of considerable importance in understanding the data and using them to answer substantive science-based questions.

As in the classification problem, the first step in clustering analysis is to define a suitable measure of distance. Since the classes are not known *a priori*, we calculate the distance between pairs of observations instead of calculating the distance between an observation and the parameters of the known classes. Suppose we have N observations. The calculation of the observation-to-observation distances leads us to $N(N-1)/2$ pair wise distances between N individuals. These are dissimilarities between individuals and should not be confused with inter-landmark distances within an individual. The dissimilarities between all individuals can be collected in an $N \times N$ matrix, called the matrix of dissimilarities where each row and each column corresponds to an individual. The dissimilarity metric calculated between individual a and individual p will be entered into the cell where column a and row p intersect and again where column p and row a intersect. The matrix of dissimilarities is a square symmetric matrix that is mathematically similar to the Euclidean Distance Matrix (Form Matrix) that was used in Chapters 3 to 5. Given such a matrix of dissimilarities, we try to construct the groups so that within a group the observations (measures of dissimilarity) are more similar than they are between groups. There are many standard statistical procedures (e.g., hierarchical clustering or k-means clustering) that may be used towards this purpose. There are also several standard statistical packages that implement these procedures. We do not discuss the details of these procedures here. WinEDMA software (http://faith.med.jhmi.edu) offers a procedure under the heading "ordination."

We feel obligated to point out that clustering is a very subjective process. First, a particular distance measure must be chosen. Second, once the analysis is run, the results of the clustering procedure are used to decide how many groups exist in the dataset. This requires that we decide what is meant by 'more' similar. These subjective choices make clustering more an art than a science. Each choice leads to different error rates and it cannot be determined that one approach is globally the best under all circumstances. Consequently, we advocate cluster analysis as a method of data exploration only. Arguments in favor or against different clustering techniques are unproductive.

6.6 Clustering analysis example

In the following example, we use the same dataset discussed in the classification section but we pretend that the classes of different individuals are not known. Thus, we mix all four datasets together and then apply clustering procedures to 'find' the groups in the aggregated dataset. If we find groups that correspond nicely with what we know, then we can say that the clustering procedure is effective in separating the different groups. On the other hand, if the clustering procedure fails to separate the groups, we say that it is not very effective. Unfortunately, real life situations do not allow the luxury of this kind of verification.

$$D_P(X_i^C, X_j^C) = tr(X_i^{C^T} X_i^C) + tr(X_j^{C^T} X_j^C)$$

$$-2tr(X_i^C X_j^{C^T} X_j^C X_i^{C^T})^{1/2}$$

As before, we describe the example analysis in terms of the Unweighted Procrustes distance. The procedure is repeated for all other dissimilarity measures considered.

Let X_1, X_2,...,X_N denote the landmark coordinate matrices corresponding to the n individuals in the combined sample.

> STEP 1: Compute the dissimilarity measure between all pairs of individuals. That is, compute
>
> $$D = \left[D_P(X_i^C, X_j^C) \right]_{i=1,2,...,n; j=1,2,...,n}$$
>
> for $i = 1,2,..., n$ and $j = 1,2,..., n$.
> STEP 2: Put these dissimilarity measures into matrix form and call that matrix the dissimilarity matrix:
> This is a square, symmetric matrix with diagonal elements equal to zero, since the dissimilarity between an individual and itself is zero.

STEP 3: Apply the Hierarchical clustering method to draw dendograms showing the affinities of various individuals. We used the Splus 2000 (MathSoft, Inc., 2000) program to conduct this analysis. The Splus function used to draw these dendograms is entitled 'hclust' and is described in Venables and Ripley (1997, 390-391). We also applied the Principal Coordinates Analysis described in Venables and Ripley (1997, 392) to study the clusters. The results are shown in Figures 6.1 and 6.2.

All methods used to form the dendograms identify the unicoronal group as distinctly separate from the other four groups. All methods also show large portions of the sagittal synostosis sample grouped together, though not all members. All four dendrograms show two rather large mixed groups made up of sagittal synosotosis, metopic synostosis, and posterior plagiocephaly cases. The original diagnosis for these disorders is made on the basis of calvarial shape (followed by computed tomography scanning), but the data used for this cluster analysis came from landmarks located on the cranial base; that part of the skull that underlies the brain (see Figure 1.5a). The reason that the cranial base landmarks were able to successfully differentiate the unicoronal cases is that it represents an asymmetric craniosynostosis; i.e., only one side of the coronal suture is closed prematurely. The premature closure of one half of the coronal suture redirects vectors of growth throughout the skull resulting in asymmetry of the cranial base (DeLeon et al., 2000). Posterior plagiocephaly, on the other hand, is a deformation that does not involve craniosynostosis and does not result in growth patterns that affect the symmetry of the cranial base. The other two categories, sagittal and metopic synostosis, involve craniosynostosis but do not produce asymmetry of the cranial base.

Although each type of craniosynostosis results in a predictable calvarial shape (Richtsmeier et al., 1999; Zumpano et al., 1999; DeLeon et al., 2001), in a previous study, we found that traditional landmarks taken on the surface of the calvarium (the intersection of suture lines) were not sufficient to differentiate sagittal synostosis patients from unaffected individuals (Richtsmeier et al., 1998) nor to completely differentiate the various types of symmetric craniosynostosis (Richtsmeier et al., 1999). What was needed for discrimination were fuzzy landmarks (see Chapter 2) that mark the location of bosses (bulges) on the surface of various bones of the calvarium. These fuzzy landmarks,

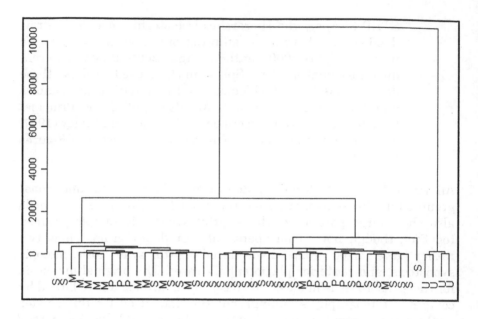

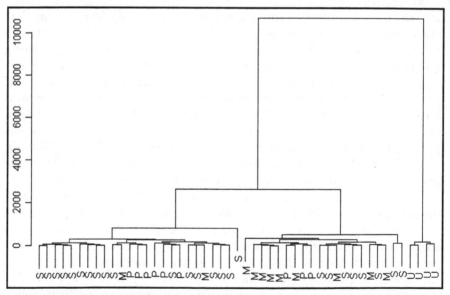

Figure 6.1. Dendograms computed by single-link clustering for various dissimilarity measures. They are in order D_P, D_A, D_R and D_{LR}. Note that all four dissimilarity measures are successful in isolating the Unicoronal synostosis cases; however, other cases do not form groups that are as clearly discernable.

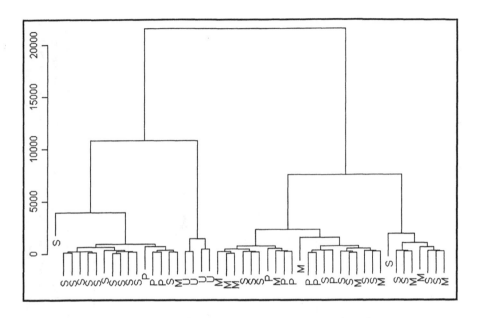

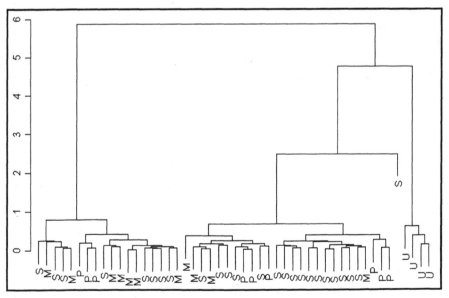

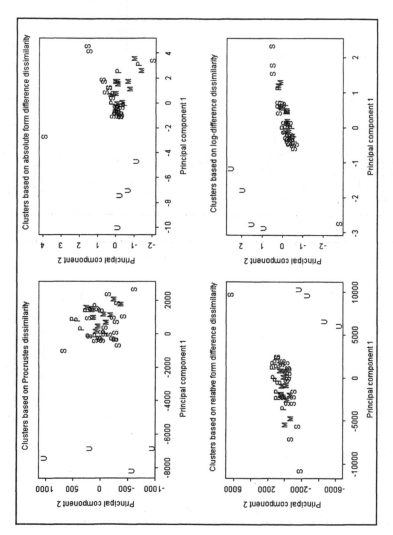

Figure 6.2 First two principal components of the dissimilarity matrix for various dissimilarity measures. Note that all four dissimilarity measures are successful in isolating the Unicoronal Synsostosis cases; however, other groups are not as clearly distinct from one another. The Procrustes dissimilarity measure indicates two groups in addition to the Unicoronal Synsostosis group. Dissimilarity measures other than the Procrustes dissimilarity measure appear to separate out some of the Sagittal synostosis cases indicating three clusters.

although more prone to measurement error (Valeri et al.,1998), contain diagnostic information on the shape of the calvarium.

This clustering analysis has demonstrated two things. First, all four methods are equally successful in isolating unicoronal synostosis using cranial base landmarks. This is due to the asymmetric nature of the cranial base in unicoronal synostosis. Second, none of the methods are able to differentiate symmetric forms of craniosynostosis and deformational posterior plagiocephaly using landmark data from the cranial base.

6.7 Summary

In this chapter, we have illustrated the use of landmark coordinate data for the purpose of classification and clustering. There are many different dissimilarity measures that may be used toward this end; we have utilized only four. By their very nature, both classification and clustering involve an arbitrary choice of a dissimilarity measure. There is no one dissimilarity measure that works best in every conceivable situation. Our studies indicate that the four different dissimilarity measures considered in this chapter can behave similarly. The reader is reminded that in another research context with alternate data, varying dissimilarity measures can provide dissimilar and even conflicting results. Our advice is to study the misclassification rate for the particular research situation using the cross-validation approach described in this chapter and choose the dissimilariy measure that works best for it. We also think that data additional to landmark coordinate data should be incorporated into the calculation of the dissimilarity measures. This additional information can potentially reduce the misclassification rate. It might also improve the performance of the clustering methods.

CHAPTER 7
Further Applications of EDMA

Theodore M. Cole III
Department of Basic Medical Science
University of Missouri-Kansas City

Dr. Tim Cole has been one of our collaborators in our work with EDMA. In this chapter, he summarizes some recent extensions of EDMA methodology to new research areas. These extensions are still under development and the discussion here is necessarily preliminary.

7.1 The study of asymmetry

Bilateral symmetry of the musculoskeletal system is a phylogenetically widespread characteristic of many complex organisms (Palmer, 1996). There is a midline plane of symmetry that divides the body into right and left halves, so that one half is essentially a mirror image of the other. However, asymmetries in form are frequently observed both in natural populations of organisms and in clinical settings. Some of these asymmetries can be very subtle. For example, while normal human faces may appear symmetric on casual inspection, some slight differences between sides can always be found. In other cases, there may be asymmetries that are immediately obvious. Some large-scale, conspicuous asymmetries may occur in normally symmetric structures because of developmental anomalies (e.g., hemifacial microsomia), and there are also dramatic asymmetries that occur normally in natural populations (e.g., the skulls of flounders).

There are three basic patterns of asymmetry, all of which are defined at the level of the sample or population: *fluctuating asymmetry*, *directional asymmetry*, and *antisymmetry* (Palmer, 1996). These patterns are distinguished on the basis of: 1) the degree of asymmetry, and 2) the "handedness" of the asymmetry. To illustrate these, let us first consider asymmetry in a single interlandmark distance. Call the right-side distance R and the left-side distance L, and collect both for a sample of organisms. For the ith individual, we can then describe any asymmetry in terms of the ratio of the right- and left-side distances: $A_i = R_i/L_i$. If ith individual is perfectly symmetric for this distance, then A_i will equal 1. If the right-side distance is larger than the left-side distance, A_i will be greater than 1. Conversely, a larger left-side distance will result in a value that is less than 1. Importantly, the distribution of A indicates the type of asymmetry that occurs in a population.

Fluctuating asymmetry: In reality, no bilaterally symmetric organism has identical right and left sides. With sufficiently precise measurement, slight differences between sides will always be found, and the handedness of these differences (indicating which side of an individual is larger) will be random. These slight differences are thought to result from the effects of environmental perturbations on normal development (Parsons, 1990; Alibert et al., 1997). Because the handedness of the fluctuations is random, the mean form for a sample is expected to be symmetric. Therefore, when an entire sample is examined, the mean of A will be 1, and the amount of dispersion will indicate the degree of asymmetry in the sample (Figure 7.1a). Traits that are more developmentally stable should exhibit smaller variances in A, while more labile traits should exhibit larger variances.

Directional asymmetry: In cases of directional asymmetry, each of the observations in a sample will be asymmetric, and there will be a handedness that characterizes the sample as a whole. In addition, directional asymmetry tends to be characterized by a greater magnitude, so that it is conspicuous (Palmer, 1996). Unlike fluctuating asymmetry, directional asymmetry is thought to have a genetic component that fixes its handedness in a population. To illustrate how directional asymmetry is expressed in the distribution of A, suppose that every observation in the sample has a right-side measurement that is substantially larger than its left-side counterpart. The result is that A will have a unimodal distribution with a mean greater than 1 (Figure 7.1b). Perhaps the most dramatic examples of directional asymmetry in natural populations are the flatfish (flounders and their relatives). While they are bilaterally symmetric as larvae, one of their eyes migrates to the opposite side of the head during metamorphosis, so that adults have both eyes on the same side of the head. In general, the handedness of the eyes is fixed within taxa, so that some species are "right-eyed" and some are "left-eyed" (although some species are

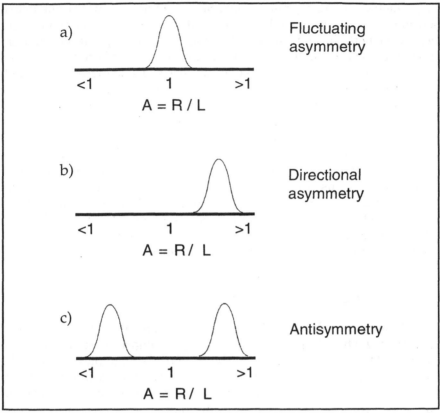

Figure 7.1 Distributions of A that are expected for different types of asymmetry. (a) Fluctuating asymmetry: Because there is no handedness in the sample, the distribution of A is unimodal and the mean is expected to be zero. (b) Directional asymmetry: The asymmetry in a typical observation is more conspicuous than in the case of fluctuating asymmetry. The distribution of A is unimodal but is not zero because of the handedness in the population. (c) Antisymmetry: The asymmetry of a typical observation is conspicuous; but there is no handedness in the population, so that the distribution of A is bimodal.

polymorphic for handedness and, therefore, are antisymmetric by definition).

Antisymmetry: Antisymmetry is similar to directional asymmetry in that each observation in a sample can be considered conspicuously asymmetric. However, there is no handedness that characterizes the sample as a whole. Each observation will have a value of A that is either greater than 1 or less than 1, so that A has a bimodal distribution (Figure 7.1c). Note that the difference between fluctuating asymmetry and antisymmetry is a matter of degree, where the distributions of A are unimodal and bimodal, respectively. Antisymmetries are seen both in clinical samples and natural populations. A well-known clinical example is hemifacial microsomia, where the bones on one side of the face are markedly undersized (compared to normal-sized bones on the opposite side). Another example is posterior plagiocephaly, where one side of the posterior neurocranium appears "flattened." In both of these examples, there is no handedness that characterizes the patient population as a whole, so that right- and left-side deformities are expected to occur with equal frequency. Antisymmetries also occur in nature, where most species of fiddler crabs in the genus *Uca* are a familiar example (Jones & George, 1982). In these taxa, each male fiddler crab has one claw that is dramatically enlarged; however, some males will have an enlarged right-hand claw and others will have an enlarged left-hand claw (so there is, again, no handedness that characterizes the population as a whole).

7.1.1. Studying Asymmetry Using EDMA

We now consider how EDMA can be applied to the analysis of asymmetry when form is measured using landmark coordinate data. In using EDMA, we can generalize the univariate examples given in the previous section to analyses where many interlandmark distances are considered. As with other EDMA applications, we are not only interested in determining *whether* a form difference occurs, but we are also interested in *where* it occurs. In other words, our studies of asymmetry aim to localize the differences in form between the left and right sides.

To quantify asymmetry patterns, we use form matrices, but we modify the way that they are constructed. In all previous applications, we described form differences between two observations or two sample means. However, in studying asymmetry, we want to measure form differences *within observations* (that is, between the right and left sides of the same observation). The algorithm for studying *directional asymmetry*, which is the most straightforward type to study, is presented below. The basics of this algorithm (where a form difference matrix is used to compare sides of the same observation) were introduced by O'Grady and Antonyshyn (1999); we have extended it to include the computation of confidence intervals.

1. Calculate sample estimates of $\hat{M}$ and Σ^*_K for the *entire object*, including all bilateral landmarks and midline points.

2. Divide the mean form into *half-forms*. The right half-form consists of all right-side bilateral landmarks and all (optional) midline landmarks. Similarly, the left half-form consists of all left-side bilateral landmarks and, again, all (optional) midline landmarks.

3. Within each half-form, calculate all distances between bilateral landmarks. Also, if midline landmarks are used, calculate the distances between each bilateral landmark and each of the midline landmarks. We do not calculate the distances between midline landmarks, as they have no bearing on the analysis. Place the right-side distances into a vector called R and the left-side distances into a vector called L.

4. Define an *asymmetry vector* (A) by elementwise division of the half-form vectors: $A = R/L$. Under the null hypothesis of bilateral symmetry, we expect all of the elements of A to be 1. Elements other than 1 indicate directional asymmetry. If an element of A is greater than 1, it indicates a distance where the right side is larger. Conversely, an element less than 1 indicates a distance where the left side is larger.

5. Compute marginal confidence intervals for the elements of A, using either nonparametric bootstrapping (Lele & Richtsmeier, 1995) or parametric bootstrapping (Lele & Cole, 1996). [These methods were introduced in Chapter 4.] If the confidence interval for an element of A contains 1, the null hypothesis of symmetry for the corresponding distance cannot be rejected. However, if the confidence interval excludes 1, the null hypothesis is rejected, and there is evidence of significant directional asymmetry.

To illustrate with real data, let us consider how EDMA can be used to study asymmetry in a clinical context. Our sample consists of eight children affected with unicoronal craniosynostosis, where the coronal suture has prematurely fused on one side of the neurocranium (Figure 7.2). The premature fusion causes marked asymmetry of the cranial vault, the cranial base, and the face. All of the children have left-side fusion, so we can treat this antisymmetry problem as a directional asymmetry problem. To make our descriptions clearer, we hereafter refer to the sides of the skull as either "fused" or "unfused" (as opposed to "left" or "right").

The landmarks describing cranial form include three pairs of bilateral landmarks and four midline landmarks. The midline landmarks are found on the face (nasion — NAS) and the basicranium (sella — SEL; basion — BAS; opisthion — OPI). The bilateral landmarks are found on the lateral rim of the orbit (frontozygomatic junction — FZJ) and on the most lateral aspect of the basicranium (external auditory meatus — EAM; asterion — AST). In the descriptions that follow, the landmark names will be prefixed with either "f", denoting the side that is prematurely fused, or "u", denoting *un*fused. To compute the asymmetry vector A, the fused-side measurements will be divided (elementwise) by the unfused-side measurements: $A = F \,/\, U$. Elements of A that are greater than 1 indicate measurements that tend to be larger on the fused side. In contrast, elements of A that are less than 1 indicate measurements that are smaller on the fused side.

The asymmetry vector (A) is shown in Table 7.1, along with 90% marginal confidence intervals that were computed using 100 nonparametric bootstrap resamples. Nearly all of the ratios are less than 1, indicating that the fused side of the cranium tends to be smaller than the unfused side (Figure 7.3). Many of these ratios have confidence intervals that exclude 1, indicating that the differences between sides are significant. This finding makes sense, because we expect premature

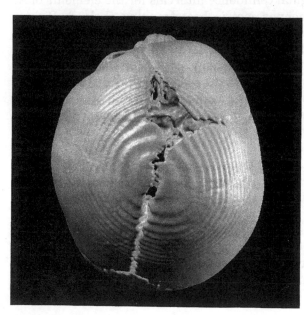

Figure 7.2 Three-dimensional CT reconstruction of a child with premature closure of the coronal suture (superior view, with anterior toward the top). The fused coronal suture is on the left side, where the frontal bone appears flattened.

suture fusion to inhibit proper growth. Two distances show more than a 10% difference between the fused and unfused sides: FZJ — EAM and FZJ — AST. This finding is expected because two measurements are perpendicular to the fused suture (FZJ is anterior to it, while EAM and AST are posterior), and the greatest limitation of growth should be in this direction. Most of the other measurements with confidence intervals excluding 1 indicate that the fused side of the cranium is shorter from front to back, which agrees with the "twisted" appearance of the cranial base and face when viewed from above (Figure 7.3).

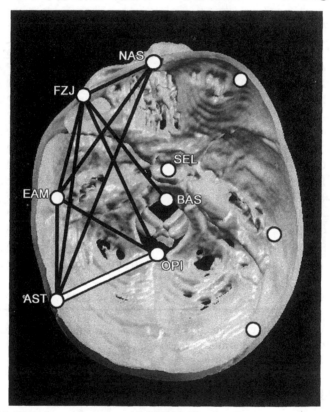

Figure 7.3 Studying asymmetry with EDMA, showing the pattern of asymmetry in the sample of unicoronal patients. The fused coronal suture is on the left side. Black lines indicate distances where the fused-side measurements are smaller than the unfused-side measurements, so that a 90% confidence interval for their ratio excludes one (see text). The white line indicates the single distance where the fused-side measurement was larger than its unfused-side counterpart. Landmarks: NAS — nasion; SEL — sella; BAS — basion; OPI — opisthion; FZJ — frontozygomatic junction; EAM — external auditory meatus; AST — asterion.

Measurements for which the null hypothesis of symmetry is not reject-ed span bilateral and midline landmarks: FZJ — SEL, EAM — BAS, EAM — SEL, AST — BAS, and AST — SEL. In general, these results indicate that the sutural fusion does little to affect the breadths of the middle and posterior cranial fossae. Finally, there is a single distance where the fused-side measurement is larger than the measurement on the unfused side (AST — OPI). This suggests that the posterior cranial fossa is broader in the fused side. However, the mean difference is only about 2% and the confidence interval barely excludes 1.

Table 7.1 Unicoronal Example

Distance	A (Fused / Unfused)	90% Confidence Interval	Pattern
FZJ — AST	0.891	0.842 — 0.938	F < U
FZJ — EAM	0.894	0.835 — 0.944	F < U
FZJ — NAS	0.919	0.864 — 0.965	F < U
EAM — AST	0.928	0.910 — 0.952	F < U
EAM — NAS	0.932	0.891 — 0.974	F < U
AST — NAS	0.932	0.896 — 0.969	F < U
FZJ — OPI	0.962	0.938 — 0.984	F < U
FZJ — BAS	0.967	0.927 — 0.999	F < U
AST — SEL	0.967	0.936 — 1.006	F = U
EAM — OPI	0.980	0.967 — 0.990	F < U
EAM — BAS	0.986	0.961 — 1.011	F = U
FZJ — SEL	0.987	0.944 — 1.021	F = U
AST — BAS	0.991	0.960 — 1.047	F = U
EAM — SEL	1.002	0.979 — 1.024	F = U
AST — OPI	1.024	1.001 — 1.047	F > U

Note: A = asymmetry vector; F = fused; U = unfused. Elements of A are sorted in ascending order.

7.2 Comparisons of molecular structures

Morphologists are only some of many scientists who are interested in comparing three-dimensional forms. Biochemists, molecular biologists, and pharmacologists have a considerable interest in describing and comparing three-dimensional macromolecular structures. This is because there is an intimate relationship between molecular structure and function, and molecules with similar structures may behave similarly in their reactions to other macromolecules in the body, to drugs, or to pesticides and other toxins. In this section, we show how EDMA can be applied to the study of molecular structures.

What are the "landmarks" used for molecular comparisons? They are atoms that are considered "homologous" for all of the molecules under consideration. To use proteins as an example, these macromolecules are made up of amino acids, and each amino acid can potentially be considered a "landmark" for structural comparisons. More specifically, we can use the spatial position of a single atom as a representative of the amino acid as a whole. The choice of this atom is arbitrary; however, it is usually the C_α atom, because all amino acids have one and only one. Note that we must assume a one-to-one correspondence between amino acids in the proteins we compare. Just as for morphological comparisons, molecules that are compared using EDMA must have the same number of *labeled* landmarks. This requirement restricts us to comparisons of molecules with similar amino-acid sequences (as in the example below), where the "homology" between amino acids can be confidently inferred. The instrumentation and quantitative methods used for estimating the three-dimensional positions of the atoms in a molecule are beyond the scope of this discussion. The reader is referred to Doucet and Weber (1996) for a review.

7.2.1 *Studying molecular structure using EDMA*

When we use EDMA to compare molecules, we can ask the same general questions that are asked in morphological comparisons: What are

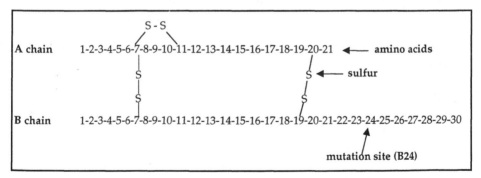

Figure 7.4 Schematic diagram of the structure of an insulin molecule (modified from Ganong, 1985). The two amino-acid chains are held together by disulfide bridges. The site of the mutation discussed in the example (B24) is indicated.

the greatest differences in form? What are the greatest similarities in form? How can these differences and similarities be localized and summarized? To demonstrate how EDMA methods can be applied to studies of molecular structure, we examine variations in insulin molecules, which are small proteins. Insulin is a pancreatic hormone that is crucial to the maintenance of normal metabolism, so that the relationship between its structure and function is of great importance (Hua et al., 1993). The structure of insulin, which is shown schematically in Figure 7.4, is very conservative across vertebrates. This conservatism helps us to feel confident about recognizing atomic homologies. There are two chains of amino acids (called the A and B chains), linked by disulfide bridges. The A chain is composed of 21 amino acids, while the B chain is composed of 30.

The data for the following examples are the three-dimensional coordinates of C_α atoms, available from the Protein Databank at Brookhaven National Laboratories (Berstein et al., 1977; Abola et al., 1987). We compare normal human insulin with a mutant form that substitutes serine (Ser) for phenylalanine (Phe) in the B24 position, denoting the 24th amino acid of the B chain (Hua et al., 1991, 1993). The amino-acid sequences are otherwise identical. The clinical consequence of this mutation — diabetes mellitus — is the same as for an insulin deficiency, even though the molecule is present in normal concentrations (Hua et al., 1993). This indicates a reduction in the bioactivity of the mutant. The amino-acid substitution changes the three-dimensional structure of the molecule, which causes a change in its function. The molecules are illustrated in Figure 7.5. Note that the atoms that describe amino acid positions are literally surrounded by empty space. There are no underlying morphological structures (as with morphological landmarks) to place the molecules in a familiar visual context for making comparisons. In addition, there is no standard coordinate system for describing molecules, so that (unlike the case with many complex organisms) terms like "anterior" and "superior" have no meaning. Finally, the large number of landmarks makes molecular comparisons even more complex.

We can use a form-difference matrix to localize the structural differences that result from the mutation. Table 7.2 shows the extremes of a sorted form-difference matrix where the mutant insulin is in the numerator and the normal insulin is in the denominator. The ten smallest values indicate distances in the mutant molecule that are much smaller than normal. Note that these distances are between amino acids that bracket the site of the mutation on the B chain (B24).

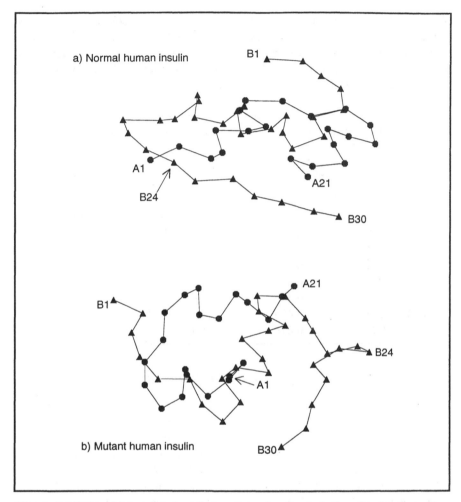

Figure 7.5 Normal (a) and mutant (b) insulin molecules. In both molecules, the positions of A- and B-chain amino acids are denoted by circles and triangles, respectively. Because there is no standard orientation for displaying them, both molecules are oriented so that their long axes are horizontal and the first molecule in the A-chain (A1) is to the left of the last (A21). The amino-acid substitution in the mutant is at position B24, producing a pronounced kink in the B chain.

Because the amino acids literally form a chain, it is useful to think of the form differences in terms of the angles between the chain's links. At B24, we can think of the angle between these segments as becoming more acute (Figure 7.5). At the other extreme of the matrix, we see that the ten largest values involve distances between the A and B chains. More specifically, we see that the distances between the mutation (B24) and the end of the A chain (A19-A21) are more than twice the normal values. Therefore, we can conclude that the mutation not only produces a kink in the B chain but also moves the terminal part of the B chain further from the A chain.

Table 7.2 Extreme values of a form difference matrix comparing mutant human insulin in the numerator and normal human insulin in the denominator.

Mutant << Normal (FDM minima)		Mutant >> Normal (FDM maxima)	
B22 — B27	0.224	B1 — B14	1.900
B21 — B26	0.380	A20 — B23	1.916
B21 — B27	0.380	A21 — B22	1.921
B22 — B26	0.398	A1 — B29	1.951
B22 — B28	0.402	A21 — B28	1.962
B23 — B27	0.446	A1 — B30	1.977
B21 — B25	0.456	A1 — B28	2.044
B22 — B29	0.462	A19 — B24	2.263
B21 — B28	0.484	A21 — B24	2.370
A19 — B21	0.514	A20 — B24	2.896

7.2.2 Similarity in shape defined by invariant cliques of landmarks

To study regions of similarity in molecules, we can use a heuristic, EDMA-based method for finding landmark "cliques." The method was originally developed for application to morphological problems, where "shape-conservative" landmark subsets are identified in a series of anatomical structures (skulls of different taxa, for example; Cole et al., 1998, 1999). However, the development of the method was inspired by algorithms that were developed in the context of molecular biology, reviewed by Willett (1991) and Doucet and Weber (1996). Those algorithms were designed to find the largest similarly shaped regions in different molecules. In the molecular biology literature, these are referred to as "largest common substructures."

Let us first describe in words what the clique-finding algorithm does. Suppose we are interested in a case where two landmark configurations are compared. Since we need at least three landmarks to describe shape, we begin by selecting a triangle as the "seed" for the first clique and then adding landmarks ("growing" the clique) through an iterative process. The seed triangle is defined as the most similarly-shaped triangle in the two configurations. Because we are interested in the aspects of shape that change the least, we need to define a statistic that measures the degree of shape difference. We can then look at all possible triangles and select the one (not necessarily unique) where the shape difference is minimum. Once the seed triangle is identified, we expand the clique by adding a landmark to the seed triangle, so that the clique has four landmarks. Which landmark should be added? We search through all possible four-landmark cliques (conditional on the three landmarks already included), and we add the landmark that creates the four-landmark clique where the shape difference is again minimal. We then add a fifth landmark using the same criterion, and so on. When do we stop adding landmarks? The clique will eventually grow to the size where the addition of any remaining landmark will result in shapes that will be too different to be considered shape-conservative. We define "too different" by specifying some *a priori* tolerance that the shape-difference statistic cannot exceed. Once the tolerance is exceeded, the clique has reached its maximum size. We then begin a search for a new clique, providing there are enough landmarks left over and that there is some remaining triangle that will meet the criterion for being shape-conservative. At most, the new clique can share only one (for two-dimensional data) or two (for three-dimensional data) landmarks with any preexisting clique. Once the second clique reaches its maximum size (beyond which the addition of any new landmark will produce configurations that are too different), the search for a new clique starts over. The analysis concludes when all possible cliques (given the tolerance) have been found.

The T statistic, defined by Lele and Richtsmeier (1991; see Chapter 4), as a coordinate-system- and scale-invariant measure of shape

$$T = \frac{\max(\mathbf{FDM})}{\min(\mathbf{FDM})}$$

difference that can be used for clique formation. Recall that and that T has a minimum possible value of 1.0, obtained when two landmark configurations have identical shapes. The *a priori* choice of a tolerance value for T is arbitrary, and we can vary the tolerance to examine its effect on how the cliques form. In many cases, a reasonable starting tolerance might be 5 to 10% of the T statistic that is observed in a form comparison that uses all of the landmarks.

To illustrate, we can again compare the structures of the mutant and normal insulins. The results of the analysis are shown in Figures 7.6 to 7.8. When all 51 landmarks are included, the T statistic is 11.85,

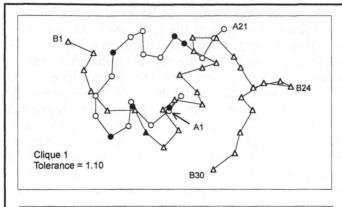

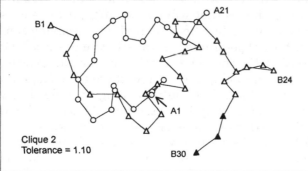

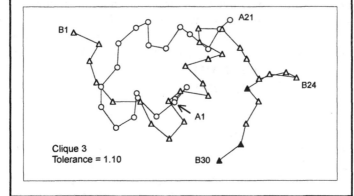

Figure 7.6
Cliques found at a low tolerance level (T=1.10), illustrated with the mutant insulin molecule. Filled symbols indicate clique membership. The amino-acid substitution in the mutant is at position B24.

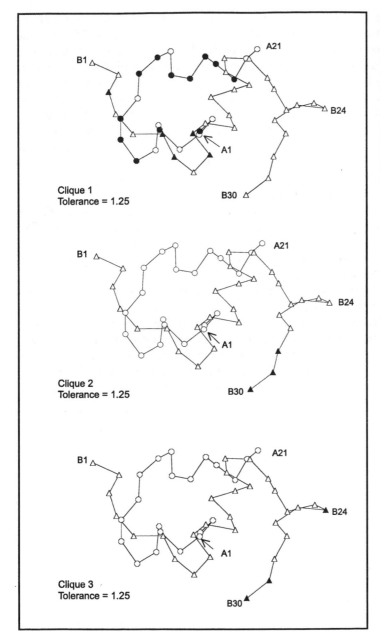

Figure 7.7
Cliques found at a medium tolerance level (T=1.25), illustrated with the mutant insulin molecule. Filled symbols indicate clique membership.

indicating a pronounced difference in overall three-dimensional structure. Beginning with a small tolerance of $T=1.10$ (Figure 7.6), we find that the most similar triangle spans the A and B chains. When the tolerance is raised to $T=1.25$ (Figure 7.7), the size of the first clique increases to include 17 amino acids (12 on the A chain and five on the B chain), shown in Figure 7.7. In addition, two new, overlapping three-landmark cliques are found at the end of the B chain: (B28, B29, and B30) and (B26, B29, and B30). This indicates that, in spite of the pronounced kink in the mutant's B chain, the structure at the end of the chain (just past B24) is largely preserved. Finally, when the tolerance is made substantially larger ($T=2.00$), we see that the clique is expanded to include all but one of the A-chain atoms (A30), as well as many of the B-chain atoms that are in close proximity to the A chain (Figure 7.8). However, little of the last third of the B chain is included. This finding complements the form-difference analysis (Table 7.2) that localized the most important form differences to the sections of the B chain that flank the mutation site.

We should note an important distinction between this algorithm and many of the common-substructure algorithms used by molecular biologists. Many of the available algorithms for molecular shape comparison do not have the requirement of specifying atom-to-atom correspondences. In many applications, the "fit" of two atoms is not a description of similarity, but is meant literally, as with the binding between a ligand and its receptor, where the molecules are so dissimi-

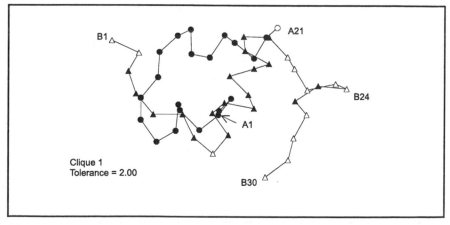

Figure 7.8 The first clique found at a high tolerance level (T=2.00), illustrated with the mutant insulin molecule. The second and third cliques are as in Figure 7.7 and are not shown.

lar in size and shape that the concept of atomic "homology" is meaningless. Therefore, many molecular algorithms are designed for use with *unlabeled* landmarks, which present a far more imposing computational problem. Because we are restricting our molecular applications to comparisons of similar molecules, where labeled atom-to-atom correspondences seem reasonable, our algorithm is much simpler to implement.

7.3 Detection of phylogenetic signals

Within the past two decades, biologists have reached a consensus regarding the importance of a phylogenetic (historical) perspective in the analysis of comparative data. This consensus holds that a phylogenetic framework is essential for addressing many questions about evolution (Coddington, 1988; Harvey and Pagel, 1991; Stearns, 1992; Rose and Lauder, 1996). While there are a number of methods available for studying univariate data, there are currently few published methods that describe *how* landmark coordinate data should be analyzed in a phylogenetic context. Part of our current research is focused on the development of such methods.

As a first step in developing new methods, we consider a very basic question: can we detect a *phylogenetic signal* in morphometric data? When we say that there is some signal in morphometric data (whether phylogenetic or not), we mean that similarities in form, growth, or shape are nonrandomly distributed across the clade. In addition, if the signal is a phylogenetic one, taxa that are closely-related will tend to be more similar to each other than they will be to more distant relatives. Put another way, we can say that morphometric data have a phylogenetic signal if our classifications based on morphometric similarities reflect genealogical relationships. To minimize the chance of making circular arguments, we use estimates of phylogeny that are based on data other than morphometrics (e.g., molecular sequence data).

Before proceeding further, we might ask why phylogenetic signals are interesting. If a phylogenetic signal does exist, it means that descendant taxa tend to inherit aspects of their form from their ancestors. Furthermore, if there are aspects of form that all members of a clade have inherited from their common ancestor, those aspects can be considered synapomorphic (= shared derived) and, therefore, homologous. In that case, we might develop hypotheses about whether those shared-derived aspects might constitute "key adaptations" for the group. Of course, if we are interested in adaptation, we might also be very interested in cases where there is a strong signal that *does not* reflect genealogy. Instead, there might be a strong "alternative" signal that reflects similarity resulting from something other than genealogy (as seen in the example below).

When we compare morphometric classifications with a genealogical hypothesis, we are comparing patterns of hierarchical relationships. The goal is to determine whether or not there is some hierarchical structure in the morphometric data. If a hierarchical structure exists for the data, we must then determine whether it matches the structure of the cladogram that is constructed from other data. To examine hierarchical structure in the data, we require a measure of dissimilarity

$$F_\Omega = \sqrt{\Sigma[\ln(\mathbf{FDM}(A,B))_{ij}]^2}$$

between taxa. One possible dissimilarity measure that we can use with landmark data is a statistic defined by Richtsmeier et al., (1998:69): where $FDM(A,B)_{ij}$ refers to all of the below-diagonal elements of the form-difference matrix between taxa A and B. If $F_\Omega = 0$, then A and B have identical forms. F_Ω will become increasingly positive as differences between taxa become more pronounced. To compare multiple taxa, all of the pairwise F_Ω statistics can be placed into a dissimilarity matrix. This matrix is then subjected to a hierarchical cluster analysis (e.g., UPGMA; Sneath and Sokal, 1973). The morphometric and cladistic trees can be compared using any of a large variety of tree-comparison and consensus statistics.

Because morphometric data vary within samples, our estimates of mean forms are always made with some sampling error. This sampling error carries through our computations of form-difference matrices, dissimilarity measures, hierarchical clustering, and tree comparisons. In other words, within-sample variation always influences our classifications that are based on morphometrics, as well as our subsequent comparisons with trees built with other data. Using the bootstrap, our knowledge of within-sample variation can be used to assess the effects of variation on our tree comparisons.

The bootstrap was first used in phylogenetic analysis by Felsenstein (1985), and most phylogenetic applications since have used the same method. The data consist of a matrix of taxa (rows) and character states (columns), which are used to generate an empirical phylogenetic tree (usually using maximum parsimony). The matrix is then resampled using a bootstrapping algorithm that selects characters (columns) randomly and with replacement. With each resampling, a bootstrap tree is constructed. When a large sample of bootstrap trees has been generated, each node in the empirical cladogram is assigned a *bootstrap proportion*. The bootstrap proportion for a node indicates the proportion of resamplings where the included taxa were grouped together in the bootstrap trees (regardless of the clades' internal structures). If there is a strong phylogenetic signal in the data, the bootstrap probabilities should be close to 100%. If there is little or no signal, the bootstrap probabilities will be closer to 0%, because the tree structure is unlikely to be reproduced during resampling.

We would like to conduct similar analyses with morphometrics. However, our application of the bootstrap is somewhat different. Felsenstein's (1985) method, and subsequent applications of it, assumes that characters are invariant within taxa. In addition, there is an assumption that the characters are independent. With morphometric characters, it is obvious that neither of these assumptions is met. In addition, we want to avoid any subjective "coding" of the data, where continuous distributions are transformed into ordinal character states. Finally, our application will differ from most in that we do not use the data to construct the cladogram; we are simply evaluating the fit between a hierarchical structure identified in our morphometric data and a cladogram that is established ahead of time, using other sources of data.

Our solution is to use a combination of parametric bootstrapping and hierarchical cluster analysis. Parametric bootstrapping assumes a model of variation, and generates random data sets under that model. Similar approaches have been used before with models of molecular sequence data (Huelsenbeck et al., 1996). We assume a general perturbation model and estimate the mean forms and variance-covariance matrices for each taxon under that model (Lele, 1993 and Chapter 3 of this book). Then, assuming multivariate normal perturbations, we can generate bootstrap samples under the model (Lele and Cole, 1996). The bootstrap samples are used to construct a sample of trees. We then compare their topologies (cluster structures) to the cladogram to see how often the genealogical relationships are reflected by the morphometric data.

As an illustration of the method, we analyzed the form of the midfacial skeleton in a sample of adult, female ateline primates (Cole et al., 2000). The atelines are a class of New World primates that are characterized by large body size (6 to 15 kg) and fully prehensile tails

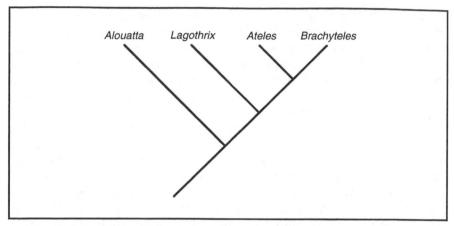

Figure 7.9 Cladogram of the living ateline primates, following Rosenberger and Strier (1989).

(Rosenberger and Strier, 1989). The living ateline genera include *Ateles* (the spider monkey), *Alouatta* (the howler monkey), *Lagothrix* (the woolly monkey), and *Brachyteles* (the muriqui or woolly spider monkey). A widely accepted hypothesis of their genealogical relationships is shown in Figure 7.9. Despite their close phylogenetic relationships, the atelines exhibit remarkable diversity in their anatomy, diet, and behavior (Rosenberger and Strier, 1989). One of the most distinctive aspects of their anatomical diversity is skull form. Six landmarks were on the facial skeletons of samples for each genus, and their mean forms and variance-covariance matrices were estimated. We then scaled each mean form according to the geometric mean of all distances, so that we could compare facial *shapes*. The dissimilarity metric described above (see also Richtsmeier et al., 1998) was used as the basis of hierarchical cluster analysis using the UPGMA criterion. The empirical clustering is shown in Figure 7.10a. Note that it is slightly different from Rosenberger and Strier's (1989) cladogram (Figures 7.9 and 7.10b). While both trees are similar in showing that *Alouatta's* shape is most distinct, the internal structure of the *Lagothrix-Ateles-Brachyteles* cluster differs. While *Ateles* and *Brachyteles* are the most closely related, *Ateles* and *Lagothrix* have the most similar shapes. When bootstrap probabilities (based on 500 resamples) are attached to both the empirical phenogram (Figure 7.10a) and the cladogram (Figure 7.10b), we see that the *Lagothrix-Ateles-Brachyteles* cluster is very stable under resampling. The phylogenetic signal for this part of the tree is very strong — *Alouatta* is not only the most distantly related, but it is also

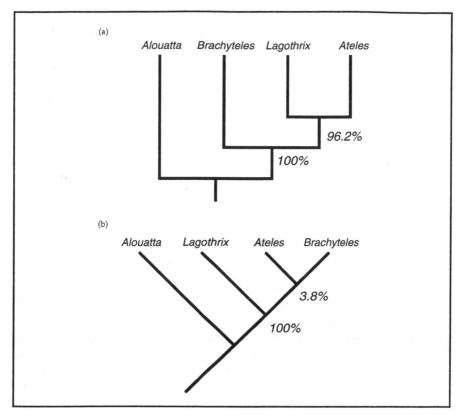

Figure 7.10 Results of the bootstrap analysis of the phylogenetic signal in the ateline data. The clustering based on morphometrics (a) differs from the cladogram (b) in the affinities of Lagothrix, Ateles, and Brachyteles. The nodes of the trees are labeled with marginal bootstrap proportions (500 resamples). The results indicate the presence of a strong "alternative" signal in the data (see text).

hypothesized to be a sister taxon of the other genera.

Within the *Lagothrix-Ateles-Brachyteles* clade, however, the phylogenetic signal is very weak. The bootstrapped morphometric data only cluster *Ateles* and *Brachyteles* together 3.8% of the time (the correct genealogy), while *Lagothrix* and *Ateles* are clustered together the remainder of the time (96.2%). Because this high bootstrap probability is associated with a difference in tree topology, it suggests a very strong signal that is *not* a phylogenetic signal, but is some kind of *alternative* signal. Cole et al. (2000) hypothesize that the strong shape similarity between *Lagothrix* and *Ateles* is due to a functional similarity related to diet, which is correlated with variations in facial structure.

Lagothrix and *Ateles* are frugivorous, while *Alouatta* and *Brachyteles* are folivorous. The primitive diet for the clade's common ancestor is believed to have been a frugivorous one (Rosenberger and Strier, 1989), so that the similarity between *Ateles* and *Lagothrix* could be considered symplesiomorphic (a retained primitive state inherited from a common ancestor). In contrast, folivory was probably developed in parallel in *Alouatta* and *Brachyteles* (Rosenberger and Strier, 1989). If that is the case, why are these genera no more similar in their facial shape? Cole et al. (2000) hypothesized that any similarities between *Alouatta* and *Brachyteles* that are due to a common diet are probably obscured by the unusually derived facial form in *Alouatta*, which is modified as the result of a hypertrophied hyolaryngeal apparatus (Rosenberger and Strier, 1989).

7.4 Summary

This chapter has offered new extensions of basic EDMA procedures as well as new algorithms for addressing complex biological problems. The examples given in the chapter demonstrate the versatile nature of the basic EDMA approach and underscore the essential importance of invariance in adapting the method to new areas of morphometric description and comparison. The applications demonstrated here are simply examples, and we are confident that future applications will be many and diverse.

Postlude

This monograph provides the foundations for quantitative analysis of landmark coordinate data based on the invariance principle. In addition to developing the statistical foundations for the study of forms and shapes as represented by landmark coordinate data, we provide descriptions and examples of various applications of our approach. The fields of application in our monograph range from Palentology, where the origins of morphometrics lay, to modern subjects of reconstructive surgery, the phenotypes of genetically engineered animal models, and molecular structure. Where should we go from here? Although we are not visionaries like D'Arcy Thompson, we' take this opportunity to speculate about the future of the field.

There is tremendous potential for the use of landmark data analysis in the fields of medicine, molecular biology, pattern recognition, computer vision, and biomechanics. One area of study of particular interest to us is the fusion of morphological data and other kinds of data, for example, behavioral genetic, life history data. This monograph discussed techniques that are most useful in exploratory or descriptive research. Now is the time we need to go beyond description and venture into explanation. To accomplish this task, we need to contemplate the construction of models for various processes that might be responsible for form change (e.g., growth, evolution, biomechanical properties, disease, genetic mutation). We hope that the next edition of this monograph will have at least a chapter explicit models for change in the geometry of biological form. an additonal aspect that requires attention is the selection and validation of models in experimental and/or natural settings. This includes validation of the Matrix Normal distribution as a sensible perturbation model as validation of explanatory models proposed by future researchers. Finally, the concept of variability its role in biological form change and the use of Σ_K^* in the study of biological variability will play a pivotal role in future applications of our methodology.

We cannot predict the future, but, for us, it has been an extraordinary journey through the morphometrics landscape. We close with the following quote that we feel particularly appropriate after almost ten years of collaboration.

So easy it seemed once found, which yet unfound most would have thought impossible.

John Milton

References

Abola, E.E., Bernstein, F.C., et al. (1987) Protein data bank. In: F.H. Allen, G. Bergerhoff, and R. Sievers (Eds.), *Data Commission of the International Union of Crystallography*, Bonn/Cambridge/Chester.

Abramowitz, M. and Stegun, I.A. (1965) *Handbook of Mathematical Functions*, New York, Dover.

Alam, K., and Mitra, A. (1990) On estimating the scale and noncentrality matrices of a Wishart distribution, *Sankhya*, Ser.B,52,133.

Alibert, P., Fel-Clair, F., et al. (1997) Developmental stability, fitness, and trait size in laboratory hybrids between European subspecies of the house mouse, *Evolution*, 51,1284.

Anstey, R. and Delmet, D. (1973) Fourier analysis of zooecial shapes in fossil Bryozoans, *Geol. Soc. Am. Bull.*, 84,1753.

Arnold, S.F. (1981) *Theory of Linear Models and Multivariate Analysis*, New York, John Wiley & Sons.

Atchley, W. (1978) Ratios, regression intercepts, and the scaling of data, *Syst. Zool.*,27,78.

Atchley, W. (1987). Developmental quantitative genetics and the evolution of ontogenies, *Evolution*, 41,316.

Atchley, W. (1993). Genetic and Developmental Aspects of Variability in the Mammalian Mandible, In: J. Hanken and B. Hall. (Eds.), *The Skull. Development*, Chicago, The University of Chicago Press.

Atchley, W., Cowley, D., et al. (1990) Correlated response in the developmental choreographies of the mouse mandible to selection for body composition, *Evolution*, 44,669.

Atchley, W., Gaskins, C., et al. (1976) Statistical properties of ratios, empirical results, *Syst. Zool.* 25,137.

Atchley, W. and Hall, B. (1991) A model for development and evolution of complex morphological structures, *Biol. Re.*, 66,101.

Barnett, S. (1990) *Matrices: Methods and Applications*, Oxford, U.K., Oxford University Press.

Barta, P., Dhingra, L., et al. (1997) Improving stereological estimates for the volume of structures identified in three-dimensional arrays of spatial data, *J. Neurosci. Meth.*, 75,111.

Bass, W. (1971) *Human Osteology: a Laboratory and Field Manual of the Human Skeleton*, Columbia, MO, The Missouri Archeological Society.

Baxter, L.L., Abrams, M., et al. (1998) Ts65Dn segmentally trisomic mice exhibit morphological changes in cerebellum similar to those in Down syndrome, *Am. J Hum. Genet.*, 63,A159.

Baxter, L.,Moran, T., et al. (2000) Discovery and genetic localization of Down syndrome cerebellar phenotypes using the Ts65Dn mouse, *Hum. Mol. Genet.*, 9,195.

Beard, L.F.H. (1967) Three-dimensional mapping by photography, *Photog. J.*,107,315.

Beard, L.F.H. and Burke, P.H. (1967) The evolution of a system of stereophotogrammetry for the study of facial morphology, *Med. Biol. Illus.* 17,20.

Behrents, R. (1985) *Growth in the Aging Craniofacial Skeleton*, Monograph 17, Ann Arbor, MI, Center for Human Growth and Development, The University of Michigan.

Benfer, R. (1975) Morphometric analysis of Cartesian coordinates of the human skull, *Am. J. Phys. Anthropol.*, 42(3),371.

Benson, R. (1982) Deformation, Da Vinci's Concept of Form and the Analysis of Events in Evolutionary History, In: E. Gallitelli (Ed.) *Paleontology, Essential of Historical Geology*, Modena, Italy, Mucchi.

Berger, J.O. (1980) *Statistical Decision Theory*, New York, Springer-Verlag.

Berstein, F.C., Koetzle, T.F., et al. (1977) The protein data bank: a computer-based archival file for macromolecular structures. *J. Mol. Biol.*,112,535.

Boas, F. (1905) The horizontal plane of the skull and the general problem of the comparison of variable forms, *Science*, 21(544),862.

Bogin, B. (1999) *Patterns of Human Growth*, 2nd ed., New York, Cambridge University Press.

Bookstein, F.L. (1978) *The Measurement of Biological Shape and Shape Change*, New York, Springer-Verlag.

Bookstein, F. L. (1982) Foundations of morphmetrics, *Annu. Rev. Ecol. Syst.* 13, 451.

Bookstein, F.L. (1986). Size and shape spaces for landmark data in two dimensions, *Stat. Sci.* 1,181.

Bookstein, F.L. (1988) Toward a notion of feature extraction for plane mappings, In: C. N. de Graaf and M. A. Viergever (Eds.) *Information Processing in Medical Imaging*, New York, Plenum Publishing.

Bookstein, F.L. (1989) Principal warps: thin-plate splines and the decomposition of deformations, *IEEE Trans. Pattern Analy. Machine Intelligence*, 11, 567.

Bookstein, F.L. (1990) Introduction to Methods for Landmark Data, *Proceedings of the Michigan Morphometrics Workshop*, F. Rohlf and F. Bookstein (Eds.), Ann Arbor, MI, The University of Michigan Museum of Zoology, 215.

Bookstein, F.L. (1991) *Morphometric Tools for Landmark Data: Geometry and Biology*, New York, Cambridge University Press.

Bookstein, F.L., Chernoff, B., et al. (Eds.) (1985) *Morphometrics in Evolutionary Biology*, Special Publication 15, Philadelphia, The Academy of Natural Sciences of Philadelphia.

Bookstein, F.L., Schafer, K., et al. (1999) Comparing frontal cranial profiles in archaic and modern *Homo* by morphometric analysis, *Anat. Rec. (New Anat.)*, 257,217.

Broadbent, B., Broadbent, B., et al. (1975) *Bolton Standards of Dentofacial Developmental Growth*, St. Louis, C.V. Mosby.

Casella, G. and Berger, R.L. (1990) *Statistical Inference*, Belmont, CA, Duxury Press.

Chapman, R. (1988). Conventional Procrustes Approaches, *Proceedings of the Michigan Morphometrics Workshop*, F. Rohlf and F. Bookstein (Eds.), Ann Arbor, MI, The University of Michigan Museum of Zoology, Special Publication No. 2, 251.

Cheverud, J. (1995) Morphological integration in the saddle-back tamarin (*Saguinus fuscicollis*) cranium, *Am. Nat.*, 145(1),63.

Cheverud, J. (1996) Developmental integration and the evolution of pleiotropy, *Am. Zool.* 36,44.

Cheverud, J., Hartman, S., et al. (1991) A quantitative genetic-analysis of localized morphology in mandibles of inbred mice using finite-element scaling analysis, *J. Craniofac. Genet. Dev. Biol.* 11,122.

Cheverud, J., Lewis, J., et al. (1983) The measurement of form and variation in form: an application of three-dimensional quantitative morphology by finite-element methods, *Am. J. Phys. Anthropol.*, 62,151.

Cheverud, J.M. and Richtsmeier, J.T. (1986) Finite-element.scaling applied to sexual dimorphism in rhesus-macaque (*Macaca mulatta*) facial growth, *Syst. Zool.*, 35,381.

Cheverud, J.M., Routman, E.J., et al. (1996) Quantitative trait loci for murine growth. *Genetics*, 142,1305.

Coddington, J.A. (1988) Cladistic tests of adaptational hypotheses, *Cladistics*, 4,3.

Cohen, M.M., Jr. (1986) *Craniosynostosis: Diagnosis, Evaluation, and Management*, New York, Raven Press.

Cole, T.M.,III (1996) Historical note: early anthropological contributions to "geometric morphometrics", *Am. J. Phys. Anthropol.*, 101(2),291.

Cole, T.M., III, DeLeon, V.B., et al. Recognition of "shape-constant" landmark cliques within larger landmark configurations, Abstract for the *1998 International Conference on Pattern Formation and Developmental Biology* (September 1998).

Cole, T.M., III, DeLeon, V.B., et al. (1999) Morphometric recognition of shape-conservative anatomical complexes, *Am. J. Phys. Anthropol.*, Supplement 280,110.

Cole, T.M., III, Lele, S., and Richtsmeier, J.T. (2000) A parametric bootstrap approach to the detection of phylogenetic signals in landmark data. (Manuscript).

Cole, T.M., III and Richtsmeier, J.T. (1998) A simple method for visualization of influential landmarks when using Euclidean Distance Matrix Analysis, *Am. J. Phys. Anthropol.*, 107,273.

Cole, T.M., III and Wall, C. (2000) Outline-based morphometrics and shape variation in the primate mandibular condyle, *Am. J. Phys. Anthropol.*, Supplement 30,127.

Corner, B. and Richtsmeier, J.T. (1991) Morphometric analysis of craniofacial growth in *Cebus apella*, *Am. J. Phys. Anthropol.*,84(3),323.

Corner, B.D., Lele, S., and Richtsmeier, J.T. (1992) Measuring precision of three-dimensional landmark data, *Quant. Anthropol.*, 3(4),347.

Coward, W. and McConathy, D. (1996) A Monte Carlo study of the inferential properties of three methods of shape comparison, *Am. J. Phys. Anthropol.*, 99,369.

Creel, N. (1978) Stereometrics in primate taxonomy and phylogeny, *Proc. Soc. Photo-Op. Instrum. Eng.*, 166,338.

Cronk, C. and Reed, R. (1981) Canalization of growth in Down syndrome children three months to six years, *Hum. Biol.*, 53(3),383.

Cuvier, G. (1828) *Le regne animal distribue e'apres son organisation*, Paris, Fortin.

Davisson, M.T., Schmidt, C., and Akeson, E. (1990) Segmental trisomy of murine chromosome 16: a new model system for studying Down syndrome, *Prog. Clin. Biol. Res.*,360,263.

Darwin, C. (1859) *On the Origin of Species*, London, John Murray.

de Gunst, M.C.M. (1987) On the distribution of general quadratic forms in normal variables, *Statistica Nierlandica*, 41(4),245.

DeLeon, V.B., Zumpano, M., and Richtsmeier, J.T. (2001) The effect of neurocranial surgery on basicranial morphology in isolated sagittal craniosynostosis, *Cleft Palate-Craniofac. J.*, in press.

DeLeon, V.B., Cole, T.M., III, et al. (2000) Reversal of asymmetric growth in the cranial base following neurocranial surgery in unicoronal craniosynostosis, *Am. J. Phys. Anthropol.*, Supplement 30,136.

Doucet, J.P. and Weber, J. (1996) *Computer-Aided Molecular Design: Theory and Applications*, London, Academic Press.

Dryden, I.L. and Mardia, K.V. (1991) General shape distributions in a plane, *Adv. Appl.Probab.*, 23,259.

Dryden, I.L., and Mardia, K.V. (1992) Size and shape-analysis of landmark data, *Biometrika*, 79,57.

Dryden, I.L. and Mardia, K.V. (1998) *Statistical Shape Analysis*, Chichester, U.K., John Wiley & Sons.

Dufresne, C. and Richtsmeier, J.T. (1995) The interaction of craniofacial dysmorphology, growth, and prediction of surgical outcome, *J. Craniofac. Surg.*, 6,270.

Efron, B. and Tibshirani, R. (1991) *An Introduction to the Bootstrap*, New York, Chapman & Hall.

Ehrlich, R., Pharr, R., et al. (1983) Comments on the validity of Fourier descriptors in systematics: a reply to Bookstein et al., *Syst. Zool.*, 32,202.

Ehrlich, R. and Weinberg, B. (1970) An exact method for the characterization of grain shape, *J. Sed. Petrol.*, 40,205.

Eveleth, P. and Tanner, J. (1976) *Worldwide Variation in Human Growth*, New York, Cambridge University Press.

Farkas, L. (Ed.) (1994) *Anthropometry*, New York, Raven Press.

Felsenstein, J. (1985) Confidence limits on phylogenies: an approach using the bootstrap, *Evolution*, 39,783.

Ferson, S., Rohlf, F., et al. (1984) A comparison of Fourier methods for the description of wing shape in mosquitoes (*Diptera: culicidae*), *Syst. Zool.*, 33,302.

Fishback, W.T. (1969) *Projective and Euclidean Geometry*, New York, John Wiley & Sons.

Fleagle, J. (1988) *Primate Adaptation and Evolution*, San Diego, Academic Press, Inc.

Friedlaender, J., Sgaramella-Zonta, L., et al. (1971) Biological divergences in South-Central Bougainville: an analysis of blood polymorphism gene frequencies and anthropometric measurements utilizing tree models, and a comparison of these variables with linguistic, geographic, and migrational "distances," *Am. J. Hum. Genet.*, 23,253.

Frostad, W., Cleall, J., et al. (1971) Craniofacial complex in the trisomy 21 syndrome (Down's syndrome), *Arch. Oral Biol.*, 16,707.

Futuyma, D.J. (1986) *Evolutionary Biology*, 2nd ed., Sunderland, MA, Sinauer Associates, Inc.

Ganong ,W.F. (1985) *Review of Medical Physiology*, 12th edition, Los Altos, CA, Lange Medical Publications.

Garza, C. and DeOnis, M. (1999) A new international growth reference for young children, *Am. J. Clin. Nutr.*, 70,1, (Part 2),169S.

Goodall, C. (1991) Procrustes methods in the statistical analysis of shape, *J. R. Stat. Soc.*, Ser. B, 53,285.

Goodall, C. and Bose, A. (1987). Models and Procrustes Methods for the Analysis of Shape Differences, *19th Symposium on the Interface between Computer Science and Statistics*.

Goodall, C. and Green, P. (1986) Quantitative analysis of surface growth, *Bot. Gaz.*, 147(1),1.

Goodwin, B. C. (1984) Changing from an Evolutionary to a Generative Paradigm in Biology, In: J. W. Polard (Ed.) *Evolutionary Theory. Paths to the Future*, Chichester, U.K., John Wiley & Sons.

Gould, S.J. (1977) *Ontogeny and Phylogeny*, Cambridge, MA, Harvard University Press.

Gould, S.J. (1981) *The Mismeasure of Man*, New York, W.W. Norton.

Gower, J.C. (1975) Generalized Procrustes analysis, *Psychometrika*, 40(1),33.

Gower, J.C. (1998) Classification, overview, *Encyclopedia of Biostatistics*, vol. 1, P. Armitage and T. Colton, (Eds.), New York, John Wiley & Sons.

Hall, B.K. (Ed.) (1994) *Homology: the Hierarchical Basis of Comparative Biology*, San Diego, Academic Press.

Hall, B.K. (1999) *Evolutionary Developmental Biology*, 2nd ed., Dordrecht, the Netherlands, Kluwer Academic Publishers.

Hall, P. and Martin, M. (1988) On the bootstrap and two-sample problems, *Aust. J. Stat.*, 30A,179.

Hanson, D., Robb, R., et al. (1997) New software toolkits for comprehensive visualization and analysis of three-dimensional multimodal biomedical images, *J. Digital Imaging*, 10,1.

Harlow, L.L., Mulaik, S.A., and Steiger, J.H. (Eds.) (1997) *What If There Were No Significance Tests?*, Mahwah, NJ, Lawrence Erlbaum Associates.

Hartman, S. E. (1986) A stereophotogrammetric analysis of occlusal morphology of extant hominoid molars, Ph.D. Dissertation, State University of New York, Stony Brook.

Harvey, P., Martin, R., et al. (1987) Life Histories in Comparative Perspective, In: B. Smuts, D. Sheney, R. Seyfarth, R. Wrangham, and T. Struhsaker (Eds.) *Primate Societies*, Chicago, The University of Chicago Press.

Harvey, P.H. and Pagel, M.D. (1991) *The Comparative Method in Evolutionary Biology*, Oxford, U.K., Oxford University Press.

Hattori, M. et al., The DNA sequence of human chromosome 21, the chromosome 21 mapping and sequencing consortium, *Nature*, 2000, 405(6784), 311.

Hua, Q.X., Shoelson, S.E., et al. (1991) Receptor binding redefined by a structural switch in a mutant human insulin, *Nature*, 354, 238.

Hua, Q.X., Shoelson, S.E., et al. (1993) Paradoxical structure and function in mutant human insulin associated with diabetes mellitus, *Proc. Nat. Acad. Sci. U.S.A.*, 90,582.

Huelsenbeck, J.P., Hillis ,D.M., and Jones, R. (1996) Parametric Bootstrapping in Molecular Phylogenetics: Applications and Performance, In: J.D. Ferraris and S.R. Palumbi (Eds.), *Molecular Zoology: Advances, Strategies, and Protocols*, New York,Wiley-Liss.

Healy-Williams, N. and Williams, D. (1981) Fourier analysis of test shape of planktonic foraminifera, *Nature*, 289,485.

Hildebolt, C. and Vannier, M. (1988) 3-D measurement accuracy of skull surface landmarks, *Am. J. Phys. Anthropol.*, 76,497.

Hildebrand, M., Bramble, D., et al. (Eds.) (1985) *Functional Vertebrate Design*, Cambridge, MA, Belknap Press of Harvard University Press.

Holtzman, D., Santucci, D., et al. (1996) Developmental abnormalities and age-related neurodegeneration in a mouse model of Down syndrome, *Proc. Nat. Acad. Sci.* U.S.A., 93,13333.

Holland, P. (1992) Discussion of the impact of sociological methodology on statistical methodology by Clifford C. Clogg, *Stat. Sci.* 7,198.

Howells, W.W. (1973) *Cranial Variation in Man. A Study by Multivariate Analysis of Patterns of Difference among Recent Human Populations*, Cambridge, MA, Peabody Museum of Archaeology and Ethnology, Harvard University.

Hrdlicka, A. (1920) *Anthropometry*, Philadelphia, Wistar Institute of Anatomy and Biology.

Huber, P. (1972) Robust statistics: a review, *Ann. Math. Stat.*, 43,1041.

Huxley, J. (1932) *Problems of Relative Growth*, London, MacVeagh.

Jacobson, A. and Sater, A. (1988) Features of embryonic development, *Development*, 104,341.

Johnson, N.L. and Kotz, S. (1970) *Distributions in Statistics: Continuous Univariate Distribution*, Vol.1, New York, Houghton Mifflin.

Jones, D.S. and George, R.W. (1982) Handedness in fiddler crabs as an aid in taxonomic grouping of the genus *Uca* (Decapoda, Ocypodidae), *Crustaceana*, 43,100.

Jungers, W. (Ed.) (1985) *Size and Scaling in Primate Biology*, New York, Plenum Press.

Kendall, D.G. (1989) A survey of the statistical theory of shape, *Stat. Sci.*, 4, 87.

Kent, J.T. (1994) The complex Bingham distribution and shape analysis, *J. R. Stat. Soc.*, Ser. B, 56,285.

Kent, J. and Mardia, K. (1997) Consistency of Procrustes estimators, *J. R. Stat. Soc.*, Ser. B 59, 281.

Kisling, E. (1966) *Cranial Morphology in Down's Syndrome: a Comparative Roentgencephalometric Study in Adult Males*, Ph.D. Thesis, Copenhagen, Munksgaard.

Kohn, L. (1994). Methods of Collecting Three-Dimensional Coordinate Data., International Congress of Vertebrate Morphology, *Morphometrics Workshop*, Chicago.

Kohn, L. and Cheverud, J. (1992) Anthropometric Imaging System Repeatability, *Electronic Imaging of the Human Body*, Dayton, OH, Cooperative Working Group in Electronic Imaging of the Human Body.

Kowalski, C.J. (1972) A commentary on the use of multivariate statistical methods in anthropometric research, *Am. J. Phys. Anthropol.*, 36,119.

Kreiborg, S., Aduss, H., et al. (1999) Cephalometric study of the Apert syndrome in adolescence and adulthood, *J. Craniofac. Genet. Dev. Biol.*,19,1.

Kreiborg, S. and Pruzansky, S. (1981) Craniofacial growth in premature craniofacial synostosis, *Scand. J. Plas. Reconstr. Surg.*, 15,171.

Krzanowski, W. (1988) *Principles of Multivariate Analysis: a User's Perspective*, Oxford, U.K., Clarendon.

Lague, M. and Jungers, W. (1999) Patterns of sexual dimorphism in the hominoid distal humerus, *J. Hum. Evol.*, 36,379.

Lauder, G.V. (1982) *Introduction. Form and Function*, E.S. Russell (Ed.), Chicago, The University of Chicago Press.

Lehmann, E.L. (1959) *Testing Statistical Hypothesis*, New York, John Wiley & Sons.

Lele, S. (1991) Some comments on coordinate-free and scale invariant methods in morphometrics, *Am. J. Phys. Anthropol.*, 85(4),407.

Lele, S. (1993) Euclidean distance matrix analysis (EDMA) of landmark data: estimation of mean form and mean form difference, *Math. Geol.*, 25(5),573.

Lele, S. (1999) Invariance and Morphometrics: a Critical Appraisal of Statistical Techniques for Landmark Data, In: M. Chaplain, G. Singh, and J. McLachlan (Eds.), *On growth and Form. Spatio-Temporal Pattern Formation in Biology*, Chichester, U.K., John Wiley & Sons, Ltd.

Lele, S. and Cole, T.M.,III. (1996) A new test for shape differences when variance-covariance matrices are unequal, *J. Hum. Evol.*, 31,193.

Lele, S. and McCulloch, C. (2000) Invariance and morphometrics, *J. Am. Stat. Assoc.* (In review).

Lele, S. and Richtsmeier, J.T. (1990) Statistical models in morphometrics: are they realistic?, *Syst. Zool.*, 39(1),60.

Lele, S. and Richtsmeier, J.T. (1991) Euclidean distance matrix analysis: a coordinate-free approach for comparing biological shapes using landmark data, *Am. J. Phys. Anthropol.*, 86,415.

Lele, S. and Richtsmeier, J.T. (1992) On comparing biological shapes: detection of influential landmarks, *Am. J. Phys. Anthropol.*, 87(1),49.

Lele, S. and Richtsmeier, J.T. (1995) Euclidian Distance Matrix Analysis: Confidence intervals for form and growth differences, *Am. J. Phys. Anthropol.*, 98,73.

Lestrel, P. (1982) A Fourier Analytic Procedure to Describe Complex Morphological Shape, In: *Factors and Mechanisms Influencing Bone Growth*, New York, Alan R. Liss, Inc.

Lestrel, P. (1989) Method for analyzing complex two-dimensional forms: elliptical Fourier functions, *Am. J. Hum. Biol.*, 1,149.

Leutenegger, W. and Cheverud, J. (1982) Correlates of sexual dimorphism in primates: ecological and size variables, *Int. J. Primatol.*, 3,387.

Lew, W. and Lewis, J. (1977) An anthropometric scaling method with application to the knee joint, *J. Biomech.*, 10,171.

Lewis, J., Lew, W., and Zimmerman, J. (1980) A nonhomogeneous anthropometric scaling method based on finite element principles, *J. Biomech.*, 13,815.

Li, Y. and Richtsmeier, J.T. (1997) Morphometric analysis of the ontogeny of sexual dimorphism of the bony pelvis in the squirrel monkey (*Saimiri sciureus, Cebidae*), *Am. J. Phys. Anthropol.*, Supplement 24,155.

Lohman, G.P. and Schweitzer, P.N. (1990) On Eigenshape Analysis, *Proceedings of the Michigan Morphometrics Workshop*, Special Publication No. 2 F.J. Rohlf and F.L. Bookstein (Eds.), Ann Arbor, MI, The University of Michigan Museum of Zoology, 147.

Lohmann, G.P. (1983) Eigenshape analysis of microfossils: a general morphometric procedure for describing changes in shape, *Math. Geol.*, 15,659.

Lozanoff, S. (1992) Accuracy and precision of computerized models of the anterior cranial base in young mice, *Anat. Rec.*, 234,618.

Lozanoff, S., Jureczek, S., et al. (1994) Anterior cranial base morphology in mice with midfacial retrusion, *Cleft Palate-Craniofac. J.*, 31,193.

Lubensky, A. and Richtsmeier, J.T. (1999) Craniofacial change in the mature and elderly: a three-dimensional comparison of sexual dimorphism in aging, *Am. J. Phys. Anthropol.* Supplement 28,187.

Ma, W. and Lozanof, S. (1996) Morphological deficiency in the prenatal anterior cranial base of midfacially retrognathic mice, *J. Anat.*, 188,547.

MacLarnon, A.M. (1989) Applications of the reflex instruments in quantitative morphology, *Folia Primatol.*, 53,33.

MacLeod, N. and Rose, K.D. (1993) Inferring locomotor behavior in Paleogene mammals via eigenshape analysis, *Am. J. Sci.*, 239-A,300.

Marcus, L.F., Bello, E., and Garica-Valdecasas., A. (Eds.) (1993) *Contributions to Morphometrics*, Madrid, Spain, Museo Nacional de Ciencias Naturales.

Marcus, L.F., Corti, M., et al. (Eds.), (1996) *Advances in Morphometrics*, NATO ASI Series A: Life Sciences, Vol. 284, New York, Plenum Press.

Mardia, K., Kent, J., et al. (1979) *Multivariate Analysis*, London, Academic Press.

McHenry, H.M. and Corrunicini, R.S., (1978) Analysis of the hominoid os coxae by Cartesian coordinates, *Am. J. Phys. Anthropol.*, 48,215.

McKinney, M. and McNamara, K.T. (1990) *Heterochrony: The Evolution of Ontogeny*, New York, Plenum Press.

McNamara, K.T. (Ed.) (1995) *Evolutionary Change and Heterochrony*, New York, John Wiley & Sons.

Medawar, P.B. (1958) *Critique of On Growth and Form by D'Arcy Wentworth Thompson*, Oxford, U.K., Oxford University Press.

Mooney, M. and Siegel, M. (1989) A test of 2 midfacial growth models using path analysis of normal human fetal material, *Cleft Palate J.*, 26,49.

Mosimann, J. (1979) Size allometry: size and shape variables with characterizations of the lognormal and generalized gamma distributions, *J. Am. Statist. Assoc.*, 63,930.

Mosimann, J. and James, F. (1979) New statistical methods for allometry with application to Florida red-winged blackbirds, *Evolution*, 33,444.

Moyers, R.E. and Bookstein, F.L. (1979) The inappropriateness of conventional cephalometrics, *Am. J. Orthod.*, 75(6),599.

Müller, G. and Wagner, G. (1996) Homology, *Hox* genes, and developmental integration, *Am. Zool.*, 36, 4.

Mullick, R., Venkataraman, S., et al. (1998) *eTDIPS: 2D/3D Image Processing System for Volume Rendering and Telemedicine*, Annual Meeting of the Society for Computer Applications in Radiology (SCAR).

Nanda, R. (1956) Cephalometric study of the human face serial roentgenograms, *Ergeb. Anat. Entwicklungsgesch.*, 75, 599.

Napier, J. and Napier, P. (1967) *A Handbook of Living Primates*, New York, Academic Press.

Neyman, J. and Scott, E. (1948) Consistent estimates based on partially consistent observations, *Econometrika*, 16,1.

O'Grady, K.F. and Antonyshyn, O.M. (1999) Facial asymmetry: three-dimensional analysis using laser scanning, *Plas. Reconstr. Surg.*, 104,928.

O'Higgins, P. (1999) Ontogeny and Phylogeny: Some Morphometric Approaches to Skeletal Growth and Evolution, In: M. Chaplain, G. Singh, and J. McLachlan (Eds.), *On Growth and Form. Spatio-Tempoal Pattern Formation in Biology*, Chichester, U.K., John Wiley & Sons, Ltd.

Oldridge, M., Zackie, E., et al. (1999) *De novo* Alu-element insertions in FGFR2 identify a distinct pathological basis for Apert syndrome, *Am. J. Hum. Genet.*, 64,446.

Owen, R. (1848) *On the Archetype and Homologies of the Vertebrate Skeleton*, London, R. and J.E. Taylor.

Palmer, A.R. (1996) From symmetry to asymmetry: phylogenetic patterns of asymmetry variation in animals and their evolutionary significance, *Proc. Nat. Acad. Sci.* U.S.A., 93,14279.

Park, W.J., Meyers, G.A., et al. (1995) Novel FGFR2 mutations in Crouzon and Jackson-Weiss syndromes show allelic heterogeneity and phenotypic variability, *Hum. Mol. Genet.* 4,1229.

Park, W.J., Theda, C., et al. (1995c) Analysis of phenotypic features and FGFR2 mutations in Apert syndrome, *Am. J. Hum. Genet.*, 57,321.

Parsons, P.A. (1990) Fluctuating asymmetry: an epigenetic measure of stress, *Biol. Rev.*, 65, 131.

Passos-Bueno, M.R., Sertie, A.L., et al. (1997) Pfeiffer mutation in an Apert patient: how wide is the spectrum of variability due to mutations in the FGFR2 gene?, *Am. J. Med. Genet.*, 71,243.

Phelps, E. (1932) A critique of the principle of the horizontal plane of the skull, *Am. J. Phys. Anthropol.* 17(1),71.

Press, W.H., Flannery, B.P, et al. (1986) *Numerical Recipies: The Art of Scientific Computing*, New York, Cambridge University Press.

Radinsky, L. (1987) *The Evolution of Vertebrate Design*, Chicago, The University of Chicago Press.

Raff, R. (1996) *The Shape of Life*, Chicago, The University of Chicago Press.

Rao, C. (1952) *Advanced Statistical Methods in Biometric Research*, New York, John Wiley & Sons.

Rao, C. (1998) Geometry of circular vectors and pattern recognition of shape of a boundary, *Proc. Nat. Acad. Sci.*, 95,12783.

Rao, C. (2000) A note on statistical analysis of shape through triangulation of landmarks, *Proc. Nat. Acad. Sci.*, 97,2998.

Rao, C. and Suryawanshi, S. (1996) Statistical analysis of shape of objects based on landmark data, *Proc. Nat. Acad. Sci. U.S.A.*, 93,12132.

Read, D. and Lestrel, P. (1986) Comment on uses of homologous-point measures in systematics: a reply to Bookstein et al., *Syst. Zool.*, 35,241.

Reardon, W., Winter, R.M., et al. (1994) Mutations in the fibroblast growth factor receptor 2 gene cause Crouzon syndrome, *Nat. Genet.*, 8,98.

Reeves, R. (1997) Exploring development and disease through germ-line genetic engineering in the mouse, *Anat. Rec.*, 253,19.

Reeves, R., Irving, N., et al. (1995) A mouse model for Down syndrome exhibits learning and behavioral deficits, *Nat. Genet.*, 11,177.

Reeves, R., Rue, E., et al. (1998) Stch maps to mouse Chromosome 16 extending the conserved synteny with human Chromosome 21, *Genomics*, 49,156.

Reichardt, C.H. and Gollob, H.F. (1997) When Confidence Intervals Should be Used instead of Statistical Significance Tests, and Vice Versa, In: L.L. Harlow, S.A. Mulaik, and J.H. Steiger (Eds.), *What If There Were No Significance Tests?*, Mahwah, NJ, Lawrence Erlbaum Associates.

Reyment, R.A. (1991) *Multidimensional Paleobiology*, New York, Pergamon Press.

Reyment, R.A., Blackith, R., and Campbell, N. (1984) *Multivariate Morphometrics*, 2nd Edition, London, Academic Press.

Rice, S. (1998) The evolution of canalization and the breaking of von Baer's laws: modeling the evolution of development with epistasis, *Evolution*, 52(3),647.

Richtsmeier, J.T., (1985) A Study of Normal and Pathological Craniofacial Morphology and Growth Using Finite Element Methods, Ph.D. Dissertation, Northwestern University.

Richtsmeier, J.T., Baxter, L., et al. (2000) Parallels of craniofacial maldevelopment in Down syndrome and Ts65Dn mice, *Devel. Dynam.*, 217,137.

Richtsmeier, J.T. and Cheverud, J. (1986) Finite element scaling analysis of normal growth of the human craniofacial complex, *J. Craniofac. Genet. Dev. Biol.*, 6(3),289.

Richtsmeier, J.T., Cheverud, J., et al. (1993) Sexual dimorphism of ontogeny in the crab-eating macaque (*Macaca fascicularis*), *J. Hum. Evol.*, 25,1.

Richtsmeier, J.T., Corner, B., Grausz, H.M., Cheverud, J., and Danahey, S. (1993) The role of postnatal growth pattern in the production of facial morphology, *Syst. Biol.*, 42(3),307.

Richtsmeier, J.T., Morris, G., et al. (1990). The Biological Implications of Varying Element Design in Finite-Element Scaling Analyses of Growth, *Annual International Conference of the IEEE Engineering in Medicine and Biology Society*, Philadelphia.

Richtsmeier, J.T, Paik, C., et al. (1995) Precision, repeatability, and validation of the localization of cranial landmarks using computed tomography scans, *Cleft Palate-Craniofac. J.* 32(3),217.

Richtsmeier, J.T., Cole, T.M., III, et al. (1998) Pre-operative morphology and development in sagittal synostosis, *J. Craniofac. Genet. Devel. Biol.*, 18(2),64.

Richtsmeier, J.T. and Lele, S. (1993) A coordinate-free approach to the analysis of growth-patterns: models and theoretical considerations, *Biol. Rev.*, 68,381.

Richtsmeier, J.T., Aldridge, K., et al. (1999) Quantification of distinct craniofacial morphologies in isolated metopic, sagittal, and unicoronal synostosis and Crouzon syndrome, using a principle coordinates application of EDMA, Presented at the *American Cleft Palate-Craniofacial Association Meetings*, Scottsdale, AZ.

Riedl, R. (1977) A systems-analytical approach to macro-evolutionary phenomena, *Q. Rev. Biol.*, 52,351.

Robb, R. and Hanson, D. (1995) The ANALYZE Software System for Visualization and Analysis in Surgery Simulation, In: S. Lavellee (Ed.), *Computer Integrated Surgery*, Cambridge, MA, MIT Press.

Rohlf, F.J. (1986) Relationships among eigenshape analysis, Fourier analysis, and analysis of coordinates, *Math. Geol.*, 18, 845.

Rohlf, F.J. (2000) Statistical power comparisons among alternative morphometric methods, *Am. J. Phys. Anthropol.*, 111,463.

Rohlf, F.J. and Bookstein, F. (Eds.) (1990), *Proceedings of the Michigan Morphometrics Workshop*, Special Publication Number 2, Ann Arbor, MI, The University of Michigan Museum of Zoology.

Rohlf, F.J. and Marcus, L.F. (1993) A revolution in morphometrics, *TREE*, 8,129.

Rohlf, F.J. and Slice, D. (1990) Extensions of the Procrustes method for the optimal superimposition of landmarks, *Syst. Zool.*, 39,40.

Rose, M.R. and Lauder, G.V. (1996) *Adaptation*, San Diego, Academic Press.

Rosenberger, A.L. and Strier, K.B. (1989) Adaptive radiation of ateline primates, *J. Hum. Evol.*, 111,717.

Rosner, B. (1995) *Fundamentals of Biostatistics*, Belmont, CA, Duxbury Press, Wadsworth Publishing Company.

Roth, V.L. (1988). The Biological Basis of Homology, In: C.J. Humphries (Ed.), *Ontogeny and Systematics*, New York, Columbia University Press.

Roth, V.L.. (1993) On Three-Dimensional Morphometrics, and on the Identification of Landmark Points, In: L.F. Marcus, E. Bello, and A. Garcia-Valdecasas (Eds.), *Contributions to Morphometrics*, Madrid, Museo Nacional de Ciencias Naturales, CSIC.

Russell, E.S. (1982) *Form and Function*, Chicago, The University of Chicago Press.

Scott, P.J. (1981) The Reflex plotters: measurement without photographs, *Photog. Rec.*, 10,435.

Searle, S.R. (1982) *Matrix Algebra Useful for Statistics*, New York, John Wiley & Sons.

Serfling, R. J. (1980) *Approximation Theorems of Mathematical Statistics*, New York, John Wiley & Sons.

Setchell, D.J. (1984) The Reflex microscope — an assessment of the accuracy of 3-dimensional measurements using a new metrological instrument, *J. Dent. Res.*, 63,493.

Shapiro, B. (1983) Down syndrome — a disruption of homeostasis, *Am. J. Med. Genet.*, 14, 241.

Shapiro, D. and Richtsmeier, J.T. (1997) Brief communication: a sample of pediatric skulls available for study, *Am. J. Phys. Anthropol.*, 103(3),415.

Siegel, M. (1979) Image recognition and reconstruction of cleft palate histological preparations: a new approach, *Cleft Palate J.* 16, 381.

Siegel, A. and Benson, R. (1982) A robust comparison of biological shapes, *Biometrics*, 38(2), 341.

Small, C. (1996) *The Statistical Theory of Shape*, New York, Springer.

Sneath, P.H.A. (1967) Trend-surface analysis of transformation grids, *J. Zool.*, London, 151, 65.

Sneath, P.H.A. and Sokal, R. (1973) *Numerical Taxonomy*, San Francisco, W.H. Freeman.

Sokal, R. and Rohlf, F. (1981) *Biometry*, New York, W.H. Freeman.

Splus 2000 (2000), Statistics Software, MathSoft, Inc., Seattle, WA.

Speculand, B., Butcher, G.W., et al. (1988) Three-dimensional measurement: the accuracy and precision of the Reflect microscope, *Br. J. Oral Maxillofac. Surg.*, 26,276.

Stearns, S.C. (1992) *The Evolution of Life Histories*, Oxford, U.K., Oxford University Press.

Stoyan, D. (1990) Estimation of distances and variances in Bookstein's landmark model, *Biometr. J.*, 32,843.

Tanner, J. (1989). *Foetus into Man: Physical Growth from Conception to Maturity*, Cambridge, MA, Harvard University Press.

Thelander, H. and Pryor, H. (1966) Abnormal patterns of growth and development in mongolism, *Clin. Pediatr.*, 5(8),493.

Thompson, D.A.W. (1992) *On Growth and Form. The Complete Revised Edition*, New York, Dover.

Ubelaker, D. H. (1989) *Human Skeletal Remains: Excavation, Analysis, Interpretation*, Washingon, D.C., Taraxacum.

Valeri, C., Cole, T.M., III, et al. (1998) Capturing data from three-dimensional surfaces using fuzzy landmarks, *Am. J. Phys. Anthropol*, 107,113

Van Valen, L. (1982) Homology and causes, *J. Morphol.*, 173,305.

Venables, W.N. and Ripley, B.D. (1997) *Modern Applied Statistics with S-Plus* 2nd. ed., New York, Springer-Verlag.

Wagner, G.P. (1989) The biological homology concept, *Annu. Rev. Ecol. Syst.*, 20,51.

Wagner, G.P. (1989) The origin of morphological characters and the biological basis of homology, *Evolution*, 43(6),1157.

Ward, R. (1989) Facial morphology as determined by anthropometry: keeping it simple, *J. Craniofac. Genet. Devel. Biol.*, 9,45.

Webster, G. (1984) The Relations of Natural Forms, In: M.W. Ho and P.T. Saunders (Eds.), *Beyond Neo-Darwinism*, London, Academic Press.

Weeks, A. and Adkins, J. (1970) *A Course in Geometry: Plane and Solid*, Lexington, MA, Ginn and Company.

Wilkie, A.O.M., Slaney, S.F., et al. (1995) Apert syndrome results from localized mutations of FGFR2 and is allelic with Crouzon syndrome, *Nat. Genet.*, 9,165.

Willett, P. (1991) *Three-Dimensional Chemical Structure Handling*, Taunton, U.K., Research Studies Press.

Wilson, D. (1999) Regular monitoring of bone age is not useful in children treated with growth hormone, *Pediatrics*, 104,5 (Part 2),1036.

Young, G. and Householder, A.S. (1938) Discussion of a set of points in terms of their mutual distances, *Psychometrika*, 3:19.

Zeger, S. and Harlow, S. (1987) Mathematical models from laws of growth to tools for biological analysis: fifty years of *Growth*, *Growth*, 51,1.

Zonneveld, F. (1987) *Computed Tomography of the Temporal Bone and Orbit*, Munich, Urban & Schwarzenberg.

Zumpano, M.P., Carson, B.S., et al. (1999) A three-dimensional morphological analysis of isolated metopic synostosis, *Anat. Rec.*, 256:1-12.

Index

A

Affine deformation, 199
Age-related groups, forming of, 220
Alouatta, 281, 283, 284
Amino acids, 272
Aneuploid mandible, mean form of, 79, 80, 81
Animal model, genetically engineered, 8
Antisymmetry, 263, 265
Apert
    sample, 137
    skull, 139
    syndrome, 138
Arithmetic form difference, 149
ASCII file, 29
Asymmetry
    handedness of, 264
    vector, 268
Ateles, 281, 282, 283
Atom-to-atom correspondences, 278
Averages, study of, 220
Avoidance, 71

B

Biological forms, foundations for study of,
    5–7
Biological organisms
    constituting group, 64
    variation among, 62
Biological variability, importance of, 6
Biology, evolutionary, 245
Biomechanical design, 2
Bivariate plots, 4
Blood pressure, reduction in, 203
Bolton-Brush growth series, 217
Bootstrap
    approach, 154, 157
    confidence intervals, 175
    procedures, 212
    proportion, 281

reference sample, 235
Brachyteles, 282, 284

C

Camera Lucida drawings, 33
Canines, 92
Captured images, landmarks from, 32
Cebus apella, 11, 230, 232
Center for Craniofacial Anomalies, 137
Centered inner product matrix, 113
Centroids, quadrilaterals superimposed on, 95
Certainty, 246
Change
    form, 132
    true form, 127
Children, neurological damage in, 14
Cholesky decomposition, 112, 120
Clade, 245
Classification
    analysis, 246
    dissimilarity measures based, 248
    methods, 247
    rule, description of, 251
Classification, clustering, and miscellaneous
    topics, 245–261
    classification analysis, 246–247
    classification analysis example, 251–254
    cluster analysis, 255–256
    clustering example analysis, 256–261
    dissimilarity measures for landmark co-
        ordinate data, 249–251
    methods of classification, 247–248
Clinic effect, 71
Cluster analysis, 255
Clustering, see Classification, clustering, and
    miscellaneous topics
Collapsing measures, 4
Column vector, multiplication of row vector
    by, 47
Common coordinate system, lack of, 134, 216

9 780367 397630